Development Misplaced

Dr. Krishan Bir Chaudhary

PARTRIDGE

A Penguin Random House Company

To order additional copies of this book, contact
Partridge India
000 800 10062 62
www.partridgepublishing.com/india
orders.india@partridgepublishing.com

Dr. GVG. Krishnamurty
(Doctor of Laws, Jhansi & London Hon. Causa)
Former Election Commissioner of India

Foreword

Since time immemorial, organized societies, whether under a king or a chief, have been eminently concerned with the field of agriculture and the role of a farmer in feeding, sustaining and protecting the populace. And so, even in modern times, agriculture and rural development occupies a prime place in a nation's progress and future. In India a significant proportion of GDP and jobs are created by agriculture. It cannot be gainsaid that about 70% of 125 crores of people are the rural poor and they depend on agriculture and fragile forests for their livelihoods.

The book "Development Misplaced" spans one of the most critical decades in the history of free India. While the process of privatization and liberalisation started a decade earlier, from 1991 onward, the decade 2001 onward saw certain decisions that opened up India's natural resources to private investors. Most importantly, our agriculture and food system was exposed to predatory capitalism with immense financial power that has reduced our farmers to a state of utter helplessness.

Modern problems and challenges include improvement of agricultural productivity, greater access to markets, promotion of assured supply of quality seeds and a thorough review of farm policies. One of the challenges is also to review disproportionate subsides to non-farm sector.

Seeds are our national heritage and are coming under threat from unethical foreign corporations who are deliberately contaminating natural seeds with alien genetic material. If our seeds are taken over by private corporations, our food security and food sovereignty, so vital for sustaining liberty, will be gone. Unfortunately, our Governments are moving in that dangerous direction.

Seeds are Nature's gift. No private corporation or individual can create seeds. India's judicial system has consistently taken the stand that the state is the trustee of the assets bestowed to us by Nature. The Government of the day has no moral right to privatize these resources.

Our farmers with a history of eight thousand years of cultivation, of conserving a mega-biological diversity of India and feeding us through the ages adequate food and nutrition are today threatened by just a handful of alien private companies with the support of some unethical scientists, bureaucrats and politicians in India.

The book focuses on the determined push of the US Government backed seeds and food multinational corporations and the dangers to our food security and food sovereignty and complete policy failure of successive Indians Governments to withstand the pressures. Divided into five parts, the first part deals with critical issues in Agriculture including farmers rights to seeds, Government's reluctance to evolve a comprehensive pro-farmer policy, spurious seeds, fertilizers and pesticides being marketed in India by American and European multinational firms, attempts of senior politicians to keep the farmers divided, undermining the competitive advantage of India's farmers by opening up agriculture sector to foreign competition without appropriate protection and framing policies that are patently against the interests of the farmers, local traders and consumers.

Protection of indigenous agriculture system and the farmers is a strategic issue as vital as national defense and security issues; perhaps more important than defense because even the best armed and trained army can't defend if good food is not available.

However, in the second section of the book that deals with policy issues we see that successive Governments since 1991 have been dismantling the protective provisions and diluting the social safety net of farmers. That process has escalated since 2009 when UPA-II came to power without the Left parties and we can see that conditions have been created for predatory global capitalism to completely dominate our food system. This is not in our national interest.

Indian Governments have long shown a fetish with signing international agreements no matter how disgraceful these are. They have brought India into the fold of World Trade Organization [WTO], when it is known that WTO serves world's most powerful food, seeds and chemicals multinational corporations, all of them based in the USA or European Union countries. This has been done despite serious opposition of a few upright senior officials who understood the nefarious objectives of WTO. The single objective of WTO is to facilitate control over world's food system by just a handful of these multinational corporations backed by the US Government and those of the European Union. This aspect has been admirable dealt with in the third section.

The fourth section specifically deals with Genetically Engineered [GE] seeds and foods. It is strange that civilized Governments of the USA and the European Union conspired to side with a handful of multinational seeds companies to patent seeds. Seeds can't be produced like toothpaste or car. I am appalled that American judiciary has not only maintained silence on the violation of farmers' right to seeds by upholding the rights of powerful seeds' corporations to collect royalty from farmers and even impose penalty for saving and replanting those seeds. Since the American judiciary has taken such decisions, there is a need to revisit Anglo-Saxon jurisprudence and identify weaknesses that have been exploited by judges to favour large corporations in violation of farmers' fundamental rights and violation of fundamental rights of people to safe foods.

I understand that genetically engineered seeds destroy the productive capacity of soil, harm human and animal health and also contaminate the natural environment irreversibly. I think it is time for Indian Judiciary to play a leading role in preventing similar attempts in India.

The last section shows that there is a huge dialectical tension within the decision making framework of the Government of India. These are actually evidences of important communications and I am happy that these have been included in the book.

This Foreword would remain incomplete if I do not refer to the eminent credentials of the author Dr. Kishan Bir Chaudhary. Apart from being the President of Bharatiya Krishak Samaj (Indian Farmer's Organisation) he has been continuously devoted to the cause and welfare of India's farmers and their contribution in feeding the nation and ensuring our health. He was also Chairman of State Farms Corporation of India and Indian Sugarcane Development Council as well Agri-Expo-95 (Farmers Participation Committee). He thus has outstanding credentials to discuss the nation's emerging scenario in the field of agriculture highlighting the challenges and as well meaningful and practical solutions.

I am happy that Dr. Krishan Bir Chaudhary brings out many vital issues of our survival as a nation and places the facts before the whole country to decide whether we are moving in the right direction or not? He has left it to the people to judge what should be done. I wish him all success in his efforts to serve our country.

Dr. GVG Krishnamurthy

22 July, 2013
New Delhi

Padam Bhusan Dr. Pushpa M Bhargava
Former Vice Chairman, National Knowledge Commission, GOI;
Former Member, National Security Advisory Board, GOI;
Former and Founder Director, Center for Cellular and Molecular Biology, Hyderabad;

Preface

This is the first book that I am aware of which discusses comprehensively all major issues in agriculture – 85 of them – and does so from the point of view of the entire farming community. Farmers produce primary food material and live in villages. Therefore, in our country, food security, farmers' security, agriculture security of the rural sector, are functionally synonymous. The book, thus, covers all of them – which means 70 per cent of India that lives in its villages.

Some of the other strong points of the book are the following:

1)   The book follows meticulously a scientific approach which implies objectivity and adherence to facts.

2)   The book touches you with its effectiveness – whether it is in regard to cash transfer policy or the water problem, land acquisition or food security policy.

3)   It suggests viable out-of-the-box solutions.

4)   It discusses linkages of agriculture with apparently disparate activities and policies – for example, with liberalization.

5)   It has a strong, fully deserved indictment of the Government and its agencies such as ICAR.

6)   One cannot but admire the plain speaking in the book. There is, thus, no camouflaging.

7)   I liked the style in the book of first making a crisp statement and then giving the supporting evidence and facts. A randomly selected example would be the five lines of the first para in section 37, which says, "The world is facing contamination of natural environment from two sources. One the uncontrollable release of radioactive contaminants from the crippled Fukushima Nuclear Power Plant and the other from illegal planting of Genetically Engineered seeds worldwide by criminal corporations like Monsanto, Bayer, DuPont, Syngenta and Dow Chemicals." Later, he says that the Japanese accident had to happen. How true! In Section 62, the title, "MNCs will dominate if Seeds Bill (is) adopted". In the second para of this Section he sums it all up in one sentence, "The Indian farmers will lose their rights on using seeds of their choice and it would mainly promote interests of the multi-national firms"

It was difficult for me to find a Section or a statement where I would not agree with whatever has been said. The book has substantially enlarged my own sphere of thinking.

The author is a familiar face on our national TV channels and is equally well-known as an incisive speaker and writer. Such a book from such a person is, indeed, most welcome. It is a must read for anyone interested in any aspect of development here or elsewhere.

PMBhargava

P M Bhargava

Acknowledgement

Globalisation of trade and economic liberalization policies pursued by India since 1991 severely impacted the agriculture sector. The incidences of farmers' suicides rose dramatically. This situation compelled me to pen some of my thoughts. I began the exercise by writing editorial comments in farmers' magazines. I also began contributing articles in several newspapers and periodicals.

I also began reflecting my views in a number of reputed national TV channels which called me for panel discussions. I was also invited to speak at a number of seminars and public meetings hosted by farmers' organizations, civil society groups and academicians. Parliamentary Standing Committees on agriculture, food and consumer affairs, petroleum and chemical fertilizers invited me to seek my views on several issues.

Many friends and colleagues of mine suggested me to compile my works in the form of a book. This book is, therefore, an humble attempt on my part.

I take this opportunity to dedicate this book to the farming community which is struggling against all odds to feed millions of hungry mouths.

I hope my efforts will open the eyes of the policymakers and the government to ponder over the issue which they have neglected so far.

My sincere thanks to the former Election Commissioner of India Dr. GVG Krishnamurty, reputed scientist and Founder Director, Centre for Cellular and Molecular Biology, Hyderabad, Padma Bhushan Dr. Pushpa M Bhargava, former Secretary, Government of India, Sh. S. P. Ghulati, Chief Promoter-Strategic Economic Management Initiative in Governance (SEMIG), Firenze- Italy, Prof. J. George, former Chairman, Commission for Agricultural Costs and Prices, Government of India, Dr. T. Haque, former Agriculture Editor of The Financial Express, Ashok B Sharma, President, Food First information and action Network-India, D. Gurusamy for their efforts and encouragements.

My most heartfelt appreciation particularly for my dearest friend Arun Srivastava, CMC, who took pains to edit this book. Also my thanks to my assistant Rahul Sharma, an IT developer for designing the book.

I also extend my sincere thanks to all my friends and colleague who stood by me in my struggle for justice and those who encouraged and helped me in making it possible for the book to see the light of the day.

Dr. Krishan Bir Chaudhary

Contents

International Issues

Dangers of Genetically Modified crops & MNC's

Some Glimpses of Author's Activities

Author on vanguard of Farmers' Interests

Major Issues in Agriculture

1

Misplaced Development causes Climate Change, Environmental & Economic Disaster

Inappropriate models for development, adoption of western lifestyle, greed for money and intensive and extensive exploitation of natural resources have pushed the country to a danger point of ecological and environmental disaster. There is no doubt our per capita energy consumption is lower than the developed nation. But that does not mean we should compete with the developed countries in Greenhouse Gas (GHG) emissions.

Climate change is a global problem and there is a need to share the responsibility collectively by the nations to reverse the trend in this climate forcing. The developed nations should voluntarily take the lead in making deeper cuts in GHG emissions in the larger interests of the humanity. India has already taken steps to formulate Climate Change Action Plan and has announced to

AUTHOR HONOURED AT INDIAN SCIENCE CONGRESS

Dr. Murli Manohar Joshi, Minister of HRD (Govt. of India) honouring **Dr. Krishan Bir Chaudhary** on 4th Jan. 2001 **Indian Science Congress** in PUSA, New Delhi

make a cut in GHG emission by 20% by 2025. Under global climate change fora, the developing countries including India enjoy "common but differentiated responsibility" that allows to carry out their development activities. But this does not mean that our development activities should be misplaced. If we follow the Western model of development and lifestyle, it would be a doom. Rather India should upgrade its traditional models of development, if necessary with appropriate scientific technology, so that our development process remains sustainable in nature. But today's tragedy is that we are hell bent on transplanting Western models for development and apeing their lifestyle. In the process the poor people of this country, particularly the farmers are facing the consequences of the misplaced policies of the Government which are most unsustainable in nature. The impact of climate change is felt on the South West Monsoon on which over 90% of the farmers depend. Low rainfall or delayed rainfall and droughts have caused problems for farmers' livelihood and agriculture. Added to this is the shrinkage in farm land due to unwarranted urbanization, setting up of large Special Economic Zones (SEZs) and transfer of prime cropped area to big corporate houses for reaping profits. These are being aided by State Government's forcible acquisition of farm & forest land and Central Government's deliberate intention of effecting a change in the central law for land acquisition.

Food security and livelihood of farmers, the giver of food security to the nation, are critically endangered in the process. The Government should realise that by following this unsustainable path of development, it is steering the country to not only environmental and ecological disaster but also to an economic doom. This unsustainable process may help in filling up the pockets of few rich, but the poor will become more poorer. The Government is happy to project a rosy picture of GDP growth, but is not concerned about raising the per capita income of the poor and the middle class of the society. The country can progress only when the per capita income of the poorest of the poor register an appreciatable rise.

Source : Editorial; Kisan Ki Awaaz; June, 2013; [http://www.kisankiawaaz.org/ Misplaced%20development%20causes%20climate%20change,%20environmental%20a nd%20economic%20disaster.html]

Save Farmlands from Corporate hawks, ensure Food Security

It is a paradoxical situation where the Government is urging for food and nutritional security by passing the Food Security Bill in the Parliament and in the same breath encouraging loss of prime farmlands. Several provisions in The Land Acquisition, Rehabilitation and Resettlement Bill, 2011 have been deliberately watered down to facilitate easy transfer of farmlands to corporate houses for reaping rich dividends.

Government should come out with a National Land Use Plan immediately and there should be fresh look at categorisation of land on basis of this Plan. India should learn lessons from Brazil which has brought big chunks of non-farm lands into agriculture. The definition of public purposes in the Land Acquisition Act 1894 is being deliberately changed to recognise activities of private industrial houses as activities for public purposes. In the name of public-private partnership and industrial projects the lands are being forcibly acquired from farmers in different parts of the country even as the land acquisition Bill is being debated in Parliament.

District Collectors should have no right to acquire land from any farmer without his consent. Public Purpose definition should be limited to core functions of the Government performed with the public money and in no case acquisition should be made for the private corporations. Any project drawing private profit cannot be considered as for public purpose. There is a shrinkage of farmland due to unwarranted rapid urbanisation signalling danger to food security. The agriculture growth has slowed down and population growth is on the rise. From where will the Government gather food to feed millions of hungry? Will it resort to imports? The global food prices are skyrocketing and imports would mean a drainage of forex earnings. On acquiring land the compensation to be paid to farmers should be based on the converted or future use of land. It should be based on highest sale for similar lands in adjacent areas, multiplied by a factor of six times (including solatium) in rural areas and four times (including solatium) in urban areas. The 10% developed land should be given to landowners on back lease without any charge. The cash amounts mentioned in the Bill should also be linked with commodity price inflation.

Dr. Krishan Bir Chaudhary addressing on Land Acquisition Bill, Mr. Harikesh Bahadur (Ex. M.P.),
Mr. Oscar Fernandes (M.P.) & Gen. Sect. AICC, Mr. T. Mainya (M.P.), Mr. P.L. Punia (M.P.)
& Chairman SC/ST Commission on 1st August 2011, Constitution Club, New Delhi

Farmers should also be permitted to do the same activities of development on their own lands on payment of external development charges as permitted to builders by the authority. Section 28A of the old Act provided for benefits to all affected families of the enhanced compensation awarded by the Court to one individual. There is no such provision in the current Bill. One should have a similar section in the new law too.

In Chapter XI of the Bill - Temporary Occupation of Land - should be deleted as it s allows Government to acquire land for a company against farmers wishes. Companies are likely to misuse the powers given under this section, as they will first take land on a temporary basis, and when it becomes totally unfit for cultivation then ask the Collector to acquire it permanently. No arbitrary powers should be given to the District Collector. To save the food security of the country any transfer of two crops agricultural land to non-agriculturists, in general, and to foreigners and NRI's , in particular, be prohibited immediately.

The Chairperson of the ruling UPA Sonia Gandhi and the Congress Vice President Rahul Gandhi are guilty of falling back on the promises they made to farmers for acquiring their land. The new amendments proposed in the Land Acquisition, Rehabilitation and Resettlement Bill 2011 are designed to grab prime farmlands at throw-a-way prices, practically alienating farmers from their very source of livelihood. Under the chairpersonship of Sonia Gandhi, the basic draft prepared by the National Advisory Council (NAC) had suggested minimum six times compensation for land acquired in rural areas and four times compensation for land acquired in urban areas on the basis of the market value. But the Bill when first introduced in the Parliament said that compensation to be paid in rural areas will be double the market value ie Indian Stamp Duty and for urban areas it will be equal to the Indian Stamp Duty plus solatium in both the cases equivalent to 100% of the market value. Further amendments has been proposed reducing the compensation amount in rural areas.

Earlier land was being acquired for public purposes which was generally meant for defence, railways, roads, irrigation, government educational institutions and hospitals. Land acquisition was never meant for setting up projects of corporates and multinational companies for reaping profits. The very definition of public purpose in the amendments has been changed to include infrastructure projects, all activities or items listed in the notification of the government in the Department of Economic Affairs – Infrastructure section vide March 27, 2012 excluding private hospitals, private educational institutions and private hotels. Practically all corporate activities in the name of public-private partnership projects, private companies for public purpose are designed to be permitted .

Earlier it was proposed in the original Bill that a committee be set up to examine the proposal for acquiring more than100 acre land . The Committee will be headed chief secretary and consist of secretaries of finance, revenue, rural development, social justice, tribal welfare, panchayati raj and concerned departments as may be specified by the State Government and three non-official experts from the relevant fields.

But in the amendments the proposal for setting up a committee has been done away with and all powers are vested in the District Collector. In the amendments only 20% of the developed land has been offered to the farmers in lieu of compensation while the UP Government in its land acquisition policy of 2011 had offered 23% of developed land which the farmers refused to accept. The Government has proposed 154 amendments changing the basic character of the original Bill which needs to be referred to the Parliamentary Standing Committee or Joint Committee or Select Committee. In the amendments the Government under section 38A (3) has proposed that no acquisition of land shall be made in the Scheduled Areas. If such acquisition takes place it shall be done only as a demonstrable last resort. Prior consent of the concerned Gram Sabha or the Panchayats or the autonomous District Councils by resolution should be obtained for acquisition of land.

If the Government thinks that the land acquisition is an urgent necessity and the interests of the farmers should be a matter of concern, it should provide the same mechanism for land acquisition all over the country as proposed for that in the Scheduled Areas. As the farmers in the country are mostly small and marginal by the size of their holdings they need the same treatment at par with their brethren in the Scheduled Areas. There should be no case for differentiation on basis of tribal and non-tribal or on basis of religion.

Source : Editorial; Kisan Ki Awaaz; April, 2013; [http://www.kisankiawaaz.org/Save%20farmlands%20from%20corporate%20hawks,%20ensure%20food%20security.html]

3

Farmers' Right over Seeds

Seeds are gifts of nature and cultural diversity. They are not a corporate invention. Passing on this ancient heritage from generation to generation is our duty and responsibility. Seeds are a common property resource to be shared for the well being of all and saved for the well being of the future generations. Hence they cannot be owned and patented.

Seed saving and sharing is religious, moral and ethical duty of every society that cannot be interfered with by any national or international law which makes seed saving and seed sharing a crime. The first right and duty of farmers is to conserve and rejuvenate biodiversity.

The conservation of biodiversity requires of necessity the saving of seeds. Laws of compulsory registration and polices for "seed replacement" undermine the freedom of farmers to save seed varieties. "Intellectual Property" laws, patent laws and breeders' rights laws violate the "law of the seed" by making it illegal to save seeds.

Farmers are inherently breeders, though their breeding objectives and methods might differ from the objectives and methods of the private seed industry. Farmers breed for diversity while the seed industry breeds for uniformity.

Farmers' breeding strategies and intellectual contribution must be recognized in order to stop the practice of using farmers' seeds as "raw material" with no intellectual contribution of farming communities. Farmers' rights arise from their past, present and future contribution to the conservation, modification and exchange of plant genetic resources. Farmers' innovation in plant breeding takes place collectively and cumulatively.

Therefore farmers' rights arising from their role as conservers and breeders have to be vested in farming communities not in individual farmers. The practice of using farmers' varieties as "raw material" to then claim patents and intellectual property rights on the basis of invention of the traits derived from farmers' varieties must be stopped.

These phenomena can be referred to as biopiracy. The global seed industry misuses the concept of "common heritage of mankind" to freely appropriate farmers' varieties, convert them into proprietary commodities and then sell them back to the same farming communities at high costs and heavy royalties. Such privatisation through patents and intellectual property violates the rights of farming communities and leads to debt, impoverishment and dispossession of small farmers.

Access to seeds and plant genetic resources must not be restricted by private property claims and patent laws, nor by withholding germplasm stored outside the region of origin. "Open Source" seeds are open pollinated varieties, which can be reproduced from year to year, generation to generation and can be saved and replanted.

The knowledge about the information embedded in seeds and germplasm is by definition not an invention but the result of cumulative collective discovery upon which additional discoveries may be based in the future.

Farmers' rights include freedom from genetic contamination and biological pollution. The introduction of untested and unknown varieties of seeds and plants must take into account the potential environmental risks as well as other potential detrimental agricultural effects.

The "Terminator" technology, better known as 'Genetic Use Restriction Technology or GURT to produce sterile, suicidal seeds that cannot reproduce is an assault on the fundamental right of farmers and the people. The introduction of such traits is designed to create a monopoly on seed and food and must be banned on a global level. Seeds embody the past and the future.

Seeds for the future have to evolve on the basis of the conservation of the widest seed diversity and crop varieties to manage the multiple challenges of food and nutritional security, food quality, climate change and sustainability.

Source : Kisan Ki Awaaz; Jan, 2012; [http://www.kisankiawaaz.org/Farmers '%20Right %20to%20Seeds.html]

4

Political will Needed to Benefit Farmers and Real Economy

Economic downturn, lack of political will in decision making, rising prices of essential commodities, corruption in public life and natural disasters like floods and drought are seriously impacting the livelihood of farmers in this country. The continuing global slowdown after the financial crisis of 2008 and the recent Eurozone crisis has seriously impacted India's economy and India cannot be isolated from its impact as we have consciously chosen to integrate with the global economy.

Hence we have to bear the burden of distress the world economy is passing through. But given the situation, no sincere efforts are made to insulate the Indian economy from the ill effects of global slowdown. Whatever the government has done so far is to protect the industry by offering them fiscal sops. The Government believes that survival of industry at any cost would help revive the economy. The Government has not yet come out of the illusion of the 'trickle down' theory.

But recently, doubting the effects of the 'trickle down' theory, it has now started talking in terms of 'inclusive growth'. However, the mindset of the Government needs to change from being a saviour of the big corporations to being the custodian of the common man by fostering growth in the real economy. It must talk in terms of 'pouring down of benefits' and not 'trickle down'. The Government has expanded the indirect tax net to the maximum extent to cover a billion people in this country. The Service Tax has been extended to cover almost all essential services. The Government's coffer is being filled by collection of indirect taxes which was paid by the common man. In fact the revenue collection from indirect taxes is much more than the collection of revenue from corporate tax [direct tax on private firms]. This justifies that the common man need to get more benefit from the Government than the corporate houses. But the situation is unfortunately different.

Not only the Corporate Houses are pampered but whatever meager benefits and subsidies are earmarked to benefit farmers and the common man are considered as charity. Some so-called intellectuals are talking in terms of the need to phase out this subsidy as it is a fiscal burden. The Government should know that the revival of the economy can take place only if it plans to foster the real economy which represents the farmers and people at the bottom of the pyramid who create real assets for the country and not mint money by gambling.

The public money mobilized by the Government by way of tax collection is filling the pockets of the corrupt people and no sincere political will is there in this country to nip the corruption in public life in the bud. The farmer and the common man are hemmed in between the rising prices of essential commodities and natural disasters like floods and drought. Floods have occurred in some parts of the country, while in major parts drought like situation is emerging on account of poor rainfall.

The cost of cultivation has gone up as farmers are resorting to the use of diesel pump set and electric power for irrigation. There is also no political will to hold the price line by cracking down on hoarders, stockists and black marketers. The prices of essential commodities are allowed to shoot up to any extent. The Corporate Houses are ready to cash in on any benefits rendered to farmers. Recently the Government hiked the Fair and Remunerative Price for sugarcane to Rs 170 a quintal. The mill owner immediately hiked the price of sugar by Rs 4 per kg.

This is totally unjustified as sugar in stock was produced from cane grown in the last kharif season when the cane prices were lower than Rs 170 a kg. Besides the system of estimating Fair and Remunerative

Price for sugarcane benefits the mill owner more than the farmers. The farmers stood to benefit when the system of estimating Statutory Minimum Price (SMP) for cane was in place.

Only a strong political will is needed to alleviate the livelihood of farmers and the common man by fostering real economy.

Source : Editorial; Kisan Ki Awaaz;   September, 2012; [http://www.kisankiawaaz.org /Political%20will%20needed%20to%20benefit%20farmers%20and%20real%20econom y.html]

5

Fetish with GDP & Techno Arrogance to Control Nature

World's total GDP, all countries included, whatever accounting systems they use, was estimated at 33-35 trillion US dollars. The value of eco-system goods and services provided free of cost by Mother Nature has been estimated at 72-74 trillion dollars.

Has it occurred to the economist Prime Minister Dr. Manmohan Singh and his techno fixers in the Planning Commission that if Nature refuses to cooperate, all his dreams of Super Power India would end up in a colossal mayhem. The nation is on a tipping point, falling into the Olduvai Gorge. Prime agricultural lands are being acquired for 'development' projects that take away nation's capacity to produce food. The latest directive of the PM actually demolishes the regulatory regime to conserve natural resources.

Deforestation goes on and on relentlessly. Even satellite data was misused and misinterpreted, showing increase in forest cover. Decline in natural vegetation is escalating erosion of top soil with each monsoonal rain. Seasonal rivers are dead; perennial rivers are becoming seasonal. Rivers originating in Uttarakhand feeding Yamuna and Ganga are so polluted that those waters are unfit for any purpose. Highly polluting industries located by these rivers are discharging untreated effluents; all have environment clearance, all have installed treatment plants but these plants work only when the inspectors arrive. Rivers get water from melting snow during summer and the seepage from the mountains during winter. These two systems make our northern rivers or Himalayan Rivers perennial. When politicians and corporations destroy the green cover to 'develop' mountain areas, it affects the great Indian farm lands of Punjab, Haryana, UP, Bihar and Bengal.

It is the shortage of water from eco-system disruption that has led farmers to depend upon underground water resources. Underground water resources are being gobbled up for exportable and commercial crops, not to ensure food and nutrition security of Indians. Forty years ago no farmer depended on diesel or electrical pump set to pump water up; today no farmer can do without that.

Production of diesel pump set adds to GDP; the gift of Nature by way of free water does not. In the globalized idiocy of multinational corporation driven agenda of valuing ecosystem goods, is an attempt to place a value on contribution of water so that private corporations can price it and charge our farmers and consumers. That process is already on. Because of the huge value of eco-system goods and services, politicians in positions of power exploit the concept of eminent domain. If a single source of globalized corruption could be identified, it is linked to untrammeled exploitation of Mother Nature's gift to mankind by misusing the power of the State.

To justify that exploitation this idiotic term GDP was coined. All economists and 'development experts' worship the GDP God. Have they ever asked a basic question of survival? Where do they get their food from? Does Walmart or Tesco produce food? They are distributors of processed food; food comes from farmlands, worked by millions of below poverty line farmers who this government believes can survive on 28 rupees per day. Try buying anything in big super stores in India for Rs 28.

World's most advanced country, United States of America, is now into record breaking longest draught. Heat has caused huge forest fires and many towns are completely gutted. The corn crop has collapsed and price has shot up to US$8.20 per bushel. Where are techno-fixers? Can even one 'expert' in the Ministry of Agriculture or in the Planning

Commission say that technology can fix Nature's power? Can the dream-boat of privatization cause rain? Can any company in the world cause rain to fall exactly on 15th June according to the Gragorian Calendar followed by Indian Meteorological Department? Rains start to fall in the month of Sawan in North Indian Plains. Sawan started on 4th July, it started to rain on 5th. It rained in Bihar and UP and in Northern Bengal, Monsoon rain.

Do politicians in Delhi who represent 1.2 billion people respect Mother Nature? Do they know that 750 million farming households depend upon the God-given weather system of India? Can economists ever assess the value of Mother Nature? Can they ever truthfully assess the contribution of Nature in GDP? In the ultimate analysis Nature ensures GDP, not techno-fixers or economists or politicians.

Source : Editorial; Kisan Ki Awaaz; August, 2012; [http://www.kisankiawaaz.org /Fetish%20with%20GDP%20and%20the%20techno-arrogance%20to%20control %20Nature.html]

6

Seed Sector in India

The seed is the most critical link in assuring food security and food sovereignty. It is a sacred code of evolution, an embodiment of life and memory, a latent world waiting to unfold. The seed comes alive in warm soil, air and moisture and germinates. Open a seed and one finds a complete baby inside. Drawing energy from the sun, it grows and multiplies manifold. Each rice seed gives hundreds of rice grains; each wheat grain gives hundreds too. Each seed and plant is unique.

Like the earth and the sky, the immense biodiversity of seeds is our collective heritage. Cumulative innovations, adaptations and selections over many generations of farming practices, the modern seeds have evolved and these seeds belong to all. They are our most vital wealth, essential for survival. Seeds are not commodities to be bought and sold at will for profiteering by monopolies.

Allowing any variety of seed or plant to become a proprietary resource is a violation of natural justice, and a great suicidal blunder of modern economic civilization. India is a global centre of origin and diversity of rice. Over 60,000 distinct rice seed varieties have been collected by Indian agricultural research centres. Many more were inter-bred in farmers' fields adapted to diverse conditions. About 19,000 rice varieties were collected by Dr Richharia alone from just Chattisgarh and

Madhya Pradesh, of which 1600 varieties were found to be high-yielding achieving over nine metric tonnes per hectare. We have a rich diversity too of wheat, millets, pulses, coarse grains, oilseeds, vegetables, tubers, fruits, spices, and medicinal plants. About 25,000 Indian varieties of dry-land crops are held by ICRISAT alone which should not have been allowed because ICRISAT represents global corporations' interests.

Dr. Krishan Bir Chaudhary inaugurating Traditional Seed Mela on 23rd June 2012, organized by S. V. University and SAARA in Tirupati (AP)

Much of our crop seed wealth has ended up in distant gene banks like the IRRI in the Philippines, CIMMYT in Mexico, or Fort Collins in USA far from its rightful owners and the cultures in which they were rooted. This wealth represents the collective bio-cultural heritage including biodiversity, food culture, ecological knowledge and value systems of local communities who freely shared and passed them down from generation to generation.

It is also the most vital resource that must be reclaimed by us to safeguard our future livelihoods and survival options especially under climate forcing and increased vulnerability of food production to erratic weather conditions. With the inevitable growing scarcity and mounting prices of non-renewable fossil fuels and chemical fertilizers, as well as growing water shortages, the High Input Variety seeds supplied by agro-industry tailored to optimal conditions are sure to face a sharp decline in yield as reports from the all over the world indicate.

Unless our farmers are able to adopt bio-diverse ecological agriculture with their own traditional, locally adapted seeds, severe food scarcity looms ahead. Today, the danger to our priceless heritage of agro-biodiversity - from proprietary commercial hybrid seeds and GM crops - is graver than ever. The GM crops threaten severe contamination of our local crop varieties through cross-pollination, as seen in the case of corn (maize) in Mexico.

The aggressive marketing of GM crops also drives local varieties out of circulation, as witnessed by the near total erosion of traditional cotton varieties in India. The creation of 'Intellectual Property Rights' (IPRs) of plant breeders over seeds and plants, especially under the "Trade Related Intellectual Properties' (TRIPs) provisions of the World Trade Organization, combined with restrictions on unregistered traditional seed varieties, is an assault on our agro-biodiversity and its free, unhindered use. Such criminalizing of the natural rights of farmers and farming communities, whose ancestors nurtured such diversity in the first place, is a mockery of natural justice.

The UPA Government's attempt of approving extremely dangerous GM crops represents a concerted thrust by agri-business to wipe out our rich heritage of agro-biodiversity. All legislations and treaties that abet the privatization of our biodiversity and our collective genetic heritage, carving out proprietary spheres for exclusive use and profiteering, must be discarded into the dustbin of history.

Our failure to do so will ultimately destroy our agriculture, millions of agricultural livelihoods, and the food and nutrition security of all.

Source : Editorial; Kisan Ki Awaaz; July, 2012;[http://www.kisankiawaaz.org/Seed %20Sector%20In%20India.html]

7

UPA Policies have failed to Solve Farmers' Problems

Agriculture as the major livelihood resource of our people and the mainstay of our economy has become an unrewarding and un-remunerative proposition. The Planning Commission is bent upon imposing the policies and technology that have no relevance to the ground reality of Indian agriculture. The Planning Commission and the Bureaucrats are misguiding the Govt. on so called advantages of policies. The right policies can put our country on the high pedestal of agricultural development.

Agricultural policies and research in India seems to have become totally incompatible to the needs of our agro-system or perhaps it has lost its goal and gone out of track. Well laid infrastructure, countrywide network of research centers and the enormous fund spent over agricultural research failed to achieve the desired goal. Major part of the funds earmarked for agricultural research is spent on establishment and very little on actual research.

They do not fit into our agro-system and have overlooked the needs of Indian agriculture, did not care to identify the real malady and suggest constructive and realistic remedial measures. The policies will make our agro-system captive at the mercy of the corporates for all time. The corporate agriculture model is not fit for our country. Their advocacy for launching a Second Green Revolution is deceitful as it is nothing but a conspiratorial ploy of making way for genetic engineering and GMO's whereas elsewhere in the world including even developed countries, these technologies are facing stiff public resistance on Bio-ethical grounds.

The agriculture policies should put on the right track .The country needs the Farmer Centric Agriculture Model and it should be based on NATURAL RESOURCE MANAGEMENT for the sustainable agriculture. The present policies and planning and research is fast moving towards a blind alley. The research, policy and planning has become a burden on the public exchequer.

It has no perceived idea, understanding and appreciation of research priorities for a predominantly agricultural country like India. The functioning of the policy makers is the inherent weakness of its leadership and inefficient management. The policy planners miserably failed to cater to the needs of Indian agriculture. It's a matter of competence & commitment and having innovative ideas as per the need of the farmers and must be fully committed towards perspective growth and development of agriculture in the country.

Source : Editorial; Kisan Ki Awaaz; July, 2011;[http://www.kisankiawaaz.org/The%20policies%20are%20failed%20to%20solve%20the%20problems%20of%20farmers.html]

8

On Seeds, GMO Authority, US - India Knowledge Initiative in Agriculture

We urge the UPA government to delete anti-farmer provisions in its proposed amendments to the Seeds Act and withdraw implementation of US-India Knowledge Initiative in Agriculture. Stop introducing the Bill for establishing NBRA. Make GEAC accountable for addressing health and environmental concerns relating to GM crops and food. We farmers are very concerned over the indecent haste in which the UPA government is acting to introduce new legislations. Given a large body of scientific evidences, GMOs will endanger our livelihood security. The UPA government should know that wining trust vote in the Parliament is not enough. It has to face the general elections due in the middle of the next year. Therefore it needs to reverse its anti-farmer policies which favours corporate houses at the expense of farmers.

On Proposed Amendments to the Seeds Act

The UPA government is planning to move an amendment to the Seeds Act in the winter session of the Parliament to give greater leverage to the corporate houses in the seeds sector. "We would like to caution the government to incorporate the views of the Parliamentary Standing Committee on Agriculture headed by Ram Gopal Yadav. I had personally appeared before the Parliamentary panel and had suggested that seeds used by farmers should not be registered," said the president of Bharatiya

Krishak Samaj, Dr Krishan Bir Chaudhary. We firmly believes that there should be only law for regulating the seed sector and the Plant Varieties Protection & Farmers' Rights Act should be the only law for this purpose. The Seeds Act and other laws should be repealed.

The Plant Varieties Protection & Farmers' Rights Act should be further strengthened in the interests of farmers. It would be a crime to hand over seed sovereignty to corporate houses. It is the duty of the Samajwadi Party which is now supporting the government and its leading MP, Ram Gopal Yadav in particular to see that the government do not move any amendment to the Seeds Act which would jeopardize the interests of farmers.

On US-India Knowledge Initiative in Agriculture

The UPA government should also withdraw from implementing the IUS-India Knowledge Initiative in Agriculture as it seeks to an upper hand to the US-based multinationals in Indian agriculture. Agri products would be opened for patent rights by US companies in the name of research. This pact is aimed at thrusting controversial technology for genetically modified (GM) crops in the country. It is strange to note that while the Opposition parties opposed tooth and nail the US-India Civilian Nuclear Deal, they did not say a word on US-India Knowledge Initiative in Agriculture which is aimed at destroying food security and livelihood security of farmers.

On the Proposed Setting Up of NBRA

The government must clarify why it is setting up the National Biotechnology Regulatory Authority (NBRA), replacing the existing regulator Genetic Engineering Approval Committee (GEAC) which is already acting as a single window clearance for biotech products. If the government feels that the GEAC is incompetent and inefficient, it should bring it to the public knowledge.

The Supreme Court, in the course of hearing a writ petition seeking a moratorium on GM crops, had ordered some improvements for introducing transparency in the functioning of GEAC. The government had always defended the functioning of GEAC in the Supreme Court. Has it got any moral right now to say that GEAC is not functioning well and needs to be replaced by NBRA? The fact is that the GEAC, without caring for any biosafety norms and transparency, has been very fast in the approval of GM crops with a view to benefit the multinational seed companies. Since 2002, GEAC approved over 175 Bt Cotton hybrids, five events and one Bt Cotton variety. It has conducted field trials of Bt. Brinjal, Bt. Okra, GM Mustard, Bt Cabbage, GM Tomato, GM Groundnut and GM Potato.

The functioning of GEAC has been questioned by many independent scientists, like the founder director of the Centre for Cellular and Molecular Biology (CCMB), Pushpa Mittra Bhargava. He called for a total review of India's experience with Bt Cotton, including how Bt technology was brought into the country. He has also sought a two to three years moratorium on GM crops, unless and until proper independent studies are done on biosafety like pollen flow, seed germination, soil microbial activity, toxicity, allergenicity, DNA finger printing, proteomics analysis, and reproductive interferences.

At the global level, independent scientists like Arpad Pusztai have questioned the safety of GM food. Pusztai has pointed out by saying "Well-designed studies, though few in number, show potentially worrisome biological effects of GM food, which the regulators have largely ignored." In India, there were reports of sheep mortality on account of grazing over Bt cotton fields in Andhra Pradesh, which the GEAC did not consider with seriousness.

There are reported cases of illegal imports of hazardous GM food, which are not approved in the country and the government has remained a mute spectator. Illegal imports of GM food are in violation of the Rules, 1989 of the Environment Protection Act, 1986. The annual amendments to the Foreign Trade Policy made in April 2006 said unlabelled GM food import would attract penal action underForeign Trade (Development and Regulation) Act, 1992. But this is not implemented in absence of guidelines.

The panel of experts and stakeholders headed by the additional director-general of National Institute of Communicable Diseases,

Shiv Lal had recommended mandatory labeling of GM food, irrespective of the threshold level. But the recommendations were not implemented either by the health ministry or GEAC. Rather, the GEAC allowed free imports of oil extracted from GM soybeans without any labeling, tests and restrictions.

The plan to set up NBRA is largely based on the recommendations of the two panels headed by MS. Swaminathan and RA Mashelkar. The suggestions made and apprehensions raised by the Indian Council of Medical Research (ICMR) in its paper - Regulatory Regime for Genetically Modified Foods :

The Way Ahead - have not been considered.

Monsanto is charging a high technology fee, which has raised the prices of Bt cotton seeds and the issue is subjudice before the MRTP Act. There are fears that pollen flow from GM crops to non-GM crops may cause problems for farmers, who may be asked to pay high technology fee for their own seeds as had been the case with the Canadian farmer Percy Schmeiser. Indian farmers, in many areas have suffered heavy losses on account of failure of Bt cotton. States like Kerala and Uttarakhand have banned GM crops and the Centre, through the NBRA, is planning to override states governments' power to regulate agriculture.

The government should make GEAC more accountable to address health and environmental concerns, rather than set up NBRA. If the government cannot ensure health and environmental safety of from the adverse effects of GM crops then there should be a moratorium on GM crops.

Source : http://fbae.org/ - 24.july.2009

9

India's Farming Crisis

The distress of farmers in India can be traced back to the introduction of technology-led, capital intensive farming in the heyday of the Green Revolution. With the advent of 'economic liberalisation' and the globalisation of trade, agrarian distress has aggravated. Unfair rules of the multilateral global trading regime have depressed global and domestic ex-farm prices, and denied Indian farmers adequate remunerative prices.

The poor farmer is squeezed between high input costs and low returns. Credit obtained from formal or informal banking systems is unable to bail him out of this precarious situation. Caught in a vicious debt trap, many farmers have resorted to suicide.

Bt Cotton and the debt trap

Misled and misguided, the changeover to Bt Cotton proved to be a total failure, causing severe losses for our cotton growers. The enormous loss and the resultant debt trap forced thousands of cotton growers to commit suicide. Bt Cotton proved to be a total fiasco. Thousands of farmers have already committed suicide, and there seems to be no end to this tragic situation. Trends in suicide remain unabated even now especially among cotton growers. Significantly, farmer suicide is reported mainly from the high-tech agriculture regions/states, such as Andhra Pradesh, Karnataka, Tamil Nadu and Punjab.

All these states have embraced capital intensive and 'cutting edge' technology in the name of boosting production. In areas where traditional agricultural practices and organic farming are prevalent, such as Orissa, Jharkhand, Bihar, Chhattisgarh, suicide is unheard of.

The ripple effect

As far as agriculture is concerned, it is alarming that India is moving towards a point of no return. From being a self-reliant nation of food surplus, the country is becoming a net importer of food. In this context, policies to promote contract and corporate farming, the use of genetically modified seeds such as Bt Cotton, and genetic synthesis in aquaculture and industrial poultry farming, threaten to undermine food security and the livelihoods of poor farmers.

There are authenticated reports that alarmingly high numbers of cattle have died from grazing on Bt Cotton residues in fodder. And, the Government of India's Genetic Engineering Approval Committee (GEAC), under the Ministry of Environment and Forestry, is baffled at the news of sheep mortality, on account of grazing in the Bt Cotton fields in Warangal district in Andhra Pradesh.

The GEAC has already admitted that toxicity studies on Bt Cotton leaves have not yet been conducted, and although it has now asked the department of biotechnology to conduct studies, the lukewarm attitude of GEAC to ascertain the level of Bt toxin responsible for killing livestock is highly questionable.

Whose trade organisation?

The production of transgenic seeds through genetic manipulation is bad science, unethical, and totally against the natural order that is responsible for the evolution and sustainability of life. It is fraught with the danger of genetic pollution and contamination, the destruction of ecosystems, environmental degradation and DNA deviations. The regimes of the WTO promote technology-driven, high-cost farming, and encourage the corporate monopolisation of the sector.

This mainly serves to promote the interests of agri-business multinationals at the expense of small and marginal farmers in developing countries. With the interests of agri-business pre-eminent, all efforts to reduce agricultural subsidies in developed countries are being stonewalled.

The link with trade liberalisation

The impact of liberalisation on agriculture is best illustrated by India's experience in the oil seed sector. The liberalisation of heavily subsidised edible oil imports led to the decline in oil seed prices in India, financially ruining oil seed growers in the country. This has totally negated previous efforts to make India almost self sufficient in the oil seed sector by 1998. Now, almost 50 per cent of edible oil is imported, resulting in annual spending of $1,800 million in foreign exchange.

Inept, amateurish and mediocre handling of our priorities is to blame for this ugly situation. The Government is playing a faulty game of cash crops over food crops, and promotion of corporate farming at the cost of traditional agro-rural systems, especially in the dry and arid zones of the country. Such policies will further add to displacement and migration to urban areas. Blind trade liberalisation and a market driven economy will throw the country into a cobweb of trans-national corporations.

Importing oil seed is only the tip of the iceberg; it is a prelude to the beginning of the end of Indian agro-systems and their ultimate take-over by multinational corporations.

Distorting or not distorting...?

Heavy subsidies given to the farming sector in developing countries are basically responsible for dismantling India's agro-system, making it economically unrewarding. Some 'trade experts' and negotiators in the developed world try to justify their misdeeds by putting subsidies into categories: trade-distorting" and "non trade-distorting."

However, all subsidies are distorting, and India needs to be emphatic about their removal in developed countries, where commitments to reduce subsidies have not been fulfilled. Instead, subsidies have been increasing, making it more difficult for developing countries to compete in the world market.

The World Bank and the IMF have become instruments in pressurising developing countries to open up their markets. The core motive of the WTO is to promote the interest of agri-business and multinationals at the expense of small, marginal and family farms across the world.   It is imperative that developing countries are given the option to apply quantitative restrictions on imports, whenever needed, to protect the livelihoods of poor farmers and agricultural workers.

India should not tolerate the obstinate and irrational attitude of developed countries, which caused the collapse of WTO negotiations. Government should remain determined in all future negotiations to focus primarily on the interests of small and marginal farmers.

Agriculture is not only for trade; it is a way of life.

Source: New Agriculturist magazine, London; March, 2007; www.new-ag.info -

10

Agrarian crisis linked with Trade Liberalization

Farmers' distress in India can be traced to the introduction of technology-led capital-intensive farming in the heyday of the Green Revolution. This situation aggravated with the advent of "economic liberalisation" and globalisation of trade. Thousands and thousands of farmers have already committed suicides and there seems to be no end to this tragic situation as the spate of suicides remains unabated even now specially among cotton growers. Suicides are reported mainly from high-tech agriculture regions such as Andhra Pradesh, Karnataka, Vidarbha in Maharashtra, Tamil Nadu and Punjab —all states which have adopted capital-intensive and the so-called cutting-edge technology in a big way in the name of boosting production. Significantly, farmers' suicides are unheard of in areas where traditional agricultural practices and organic farming are prevalent, such as Orissa, Jharkhand, Bihar, Chhattisgarh and the northeast.

How did this situation arise? Unfair rules of the multilateral global trading regime have depressed global and domestic prices and denied the Indian farmer adequate remunerative prices. The poor farmer is squeezed between high input costs and low returns. Any amount of credit extended from the formal or informal banking system would be unable to bail him out of this precarious situation. Caught in a vicious debt trap, many farmers have taken to suicide.

It is alarming that India is moving towards a point of no return as far as Agriculture is concerned. From being a self-reliant and food surplus nation, the country is being pushed into becoming a net food-importing nation. In this context, the policy to promote contract and corporate farming, use of genetically modified seeds like Bt. Cotton, genetic synthesis in the sphere of aquaculture and industrial poultry farming are threatening to undermine food security and livelihood concerns of poor farmers. Mislead and misguided, the change over to Bt. Cotton proved to be a total failure causing severe financial loss to our cotton growers.

The rampaging enormous loss and the resultant debt trap forced thousands of cotton growers to commit suicide. Bt cotton proved to be a total fiasco. There are authenticated reports of alarmingly large number of deaths of Cattles on account of grazing of residue parts of Bt. Cotton as fodder. It is very sure that the hazardous toxin developed by Bt. genes in the Cotton plant was responsible for causing death to the Cattles.

Govt. of India's, Genetic Engineering Approval Committee (GEAC) under the Ministry of Environment & Forest is baffled at the news of sheep mortality on account of grazing over Bt cotton fields in Warangal district in Andhra Pradesh. It has asked the department of biotechnology to conduct toxicity studies on Bt cotton leaves. GEAC has already admitted that toxicity studies on Bt cotton leaves have so far not been conducted.

The reported cases of sheep mortality have become a major roadblock in the process of approval of Bt Brinjal for large scale field trials. The lukewarm attitude of GEAC to ascertain the level of Bt toxin responsible for killing and endangering the life of sheep and other grazing cattle is highly questionable. By aimlessly caricaturing genetic engineering and thrusting upon our agro system, even though the world over, production of transgenic seeds through genetic manipulation is being termed as bad science. Genetic manipulations are unethical and totally against the nature responsible for the evolution and sustainability of life and are fraught with the danger of genetic pollution and contamination of life forms, destruction of the ecosystem on earth, environmental degredation and DNA aberrations.

The WTO regime promotes technology-driven high-cost farming and encourages monopoly of corporate houses in the sector. This mainly serves to promote the interests of agri-business multinationals at the expense of small and marginal farmers of the developing countries. Keeping the interests of their agri-business in view, developed countries are stonewalling all efforts aimed at reducing agricultural subsidies. Impact of liberalization of agriculture imports is best illustrated by the experience in edible oils. The oil seeds sector in India has suffered a terrible blow on account of liberalisation of oil imports. Indiscriminate liberalization of the policy of heavily subsidized edible oil imports led to decline of oil seeds prices, financially ruining the oil seeds growers in the country.

More than 50% of the total consumption of edible oil is being imported resulting in an annual outgo of $1,800 million in foreign exchange. It totally negated and nullified the efforts made in this direction as India achieved near self-sufficiency in the oil seeds sector by 1998. In the larger national perspective the import of wheat is equally damaging. The current policy of import of wheat has exposed our wheat growers to meet the same fate as the oil seeds growers. Our overflowing grain silos and warehouses were actually creating glut and storage problems. Now, the government is crying wolf, simply for making out a case for import of wheat. It will add to further destabilization of rural economy. The decision to import wheat is highly outrageous and prejudicial to the interest of our farmers. Domestic pre-requisites have been given a go-by on extraneous considerations.

This will obviously benefit the market forces and the multinationals like CARGILL and ITC. Post W.T.O, our dream to become a global agricultural entity has been stone walled. Inept, amateurish and mediocre handling of our priorities is to blame for this ugly situation. The government is playing a foolish game of giving precedence to cash crop over food crops and promotion of corporate farming at the cost of traditional agro-rural system, especially in the dry and arid zone of the country. This will further add to displacement and migration to cities and urban belt. Food situation even after five decades of independence is gloomy and the per capita availability of food grains has rather gone down. Blind trade liberalisation and market driven economy will throw the country into the cobweb of transnational corporations. Import of wheat is only the tip of iceberg. It is a prelude to the beginning of the end of Indian agro-system and its ultimate take over by MNCs.

Heavy subsidies given to their farm sector by the developed countries are basically responsible for gradually dismantling India's agro-system and making it economically unrewarding. India needs to be very emphatic on immediate and effective removal of subsidies in developed countries. Notwithstanding the fact that all subsidies are trade distorting, some so-called trade experts and negotiators of the developed world try to justify their misdeeds by categorising subsidies and support as "trade-distorting" and "non trade-distorting". Even WTO sings to this tune of categorisation.

The developed countries have not fulfilled their commitments to reduce their subsidies and supports in the first phase of the Agreement on Agriculture (AoA). Instead, they have increased subsidies by 50% by shifting it from one box to another. In the face of increasing subsidies, it has become very difficult for developing countries to compete in the world market. Unequal globalization has hampered poverty alleviation and severely endangered India's food security. Provision of special products and special safeguard mechanism are just inadequate to safeguard the livelihood of Indian farmers. It is imperative that developing countries be given the option to apply quantitative restrictions on imports, whenever needed, to protect the livelihood of poor farmers and agricultural workers. Notwithstanding the obstinate and irrational attitude of developed countries causing collapse of WTO negotiations, India should remain fully determined in all future negotiation to focus primarily to secure and safeguard the interest of small and marginal farmers.

WTO's core motive is to promote the interest of agri-business multinationals at the expense of small and marginal farmers and family farms across the world. Agriculture negotiations should be from the point of view of the farmers and not from the point of view of multinationals. Agriculture is not only for trade, it's a way of life of the people in the developing countries. The World Bank and the IMF have become instruments in pressurizing developing countries to open-up their markets.

Now a complete about turn by the Planning Commission

The process is on in the Planning Commission for formulation of the document for the 11th Plan (2007-2012). This has given an opportunity to policy-makers and economists for projecting rosy targets for farm growth. The often quoted ambitious target for annual farm growth is 4%, which the economists say is necessary to push the overall GDP growth rate to 8% and above. The economists have rightly realised the need for boosting farm growth if the magic figure of overall GDP growth is to be met. A large section of the population is dependent on rural economy and as much as 70% of them earn their livelihood in rural areas, despite temporary migration to urban areas.

Most of the traditional village economic activities have been eroded with the advent of the new economic order. Only agriculture, animal husbandry and poultry have remained as the principal source of livelihood in rural areas. It is interesting to note a confession made by the Planning Commission on the ambitious growth targets in agriculture being projected by the economists. In an approach paper to the 11th five-year plan it said; "Accelerating GDP growth in agriculture to around 4% as envisaged in this paper, is not an easy task. Actual growth of agricultural GDP, including forestry and fishing, was only 1% per annum in the first three years of the 10th Plan and even the rosiest projections for 2005-06 and 2006-07 would limit this below 2% for the full five-year period. Several modelling exercises suggest that 4% growth of agriculture will not be sustainable from the demand side unless aggregate GDP growth is much higher than 8%." The approach paper also mentions about the supply side constraints.

The Planning Commission has presented an alternate way of push up the farm growth rate, missing real issue which has compelled farmers to commit suicides. But this amounts to escaping from the reality of the situation. Another admission made by the Plan panel helps to make the situation clearer. It said: "The supply side challenge of doubling agricultural growth is also formidable. This is especially so because no dramatic technological breakthrough comparable to the green revolution is presently in sight." Recently the government said that it has plans to increase production of wheat by 5 to 7 million tonnes. Similar plan is on the anvil for boosting production of other winter crops as well. This can only happen through area expansion under specific crops. But total cultivable land in the country is limited and is rather shrinking due to rapid urbanisation and acquisition of prime farmlands in the name of so-called development projects.

Shift in area can happen from one crop to another and that too in not all cases as each specific crop is suited to a particular agro-climatic conditions. If the policymakers are sincere in boosting farm production, then the first priority should be given for conservation of farmlands. The Planning Commission has made an honest admission by saying that no dramatic technological breakthrough in agriculture is in sight. The claims of transgenic crops boosting farm output have proved to be a hoax. In fact, transgenic crop developed across the world so far are for herbicide and insect tolerance and not for boosting productivity in real sense.

What the Planning Commission ignored is the fallout of the chemical agriculture under green revolution. Green revolution, no doubt, caused a spurt in production in its initial phase which later stagnated in its growth. The reason for this is the sharp decline in factor productivity. Farmers who have revisited organic farming have begun reaping the benefits of increased production.

Corporate Farming: Sabotaging a System

The smoke and mirror will not hide the ugly reality of the sinister design of the agri-business multinationals and corporate houses to cast their gripping hold on India's farmland. Colossal and enormous loss that will be inflicted by contract or corporate farming on our traditional agro-system, social-cultural ethos, growth factors, biodiversity and the eco-system is beyond comprehension, making it absolutely difficult to reverse the situation. Sabotaging and dismantling a system that sustained the life and times of a vast majority of our polity and provided sustenance to the agro-rural segment of our society for centuries will usher India into a state of economic subjugation.

Alluring unsuspecting simple farmers by eulogizing corporate farming as the panacea for their socio-economic emancipation, the corporate sector will fasten their grip on India's farmland, turning the source of sustenance and livelihood for the peasantry into a market driven entity. The entire scheme is outrageous and totally unacceptable. Farmers leasing out their land on contract to the corporate sector will get reduced to the status of bonded wage earner in their own farm. Small and marginal farmers having less than two acres of land holding, facing various natural and man-made constraints, still manage to sustain their livelihood from their family farms.

It is highly erroneous to think that market driven commercial venture like corporate farming can ever be trusted as a dependable source of a rewarding venture to the farmers. Diabolical policy on the issue will further debase the agro-rural segment. The situation in the countryside is already extremely dismal and the policy back-up to promote corporate farming which is not at all compatible to our traditional farming pattern, will in course of time devastate a system that is still alive and contribute significantly to country's economic growth and development.

Growth and development of agriculture should be all about sustainability, self sufficiency and self-reliance. But policies being pursued defy all logic and reflect cynicism that despite abundance of resources, India has been destined to become a food importing nation. Something seems to have seriously gone wrong. There is urgent need for the policy makers to peep into the causes of failures and re-fixing of priorities by bringing agriculture in the centre stage of economic growth and development, otherwise the way we are moving, nothing can check us from becoming a satellite nation.

Just a very pertinent question that must be put to those who are not the least-concerned over the growing command of multinational agri-business companies on seed trade in the country. Seed the basic input but exorbitant price charged by MNCs has made it unaffordable and inaccessible for our small and marginal farmers. But perhaps revitalization and reactivation of public sector seed companies are not the priorities of our policy makers. Corporate farming howsoever attractive it may appear is ultimately prove to be counterproductive and harm the farmers and the consumers alike. It is nothing but sugar coated sweet poison.

Corporate farming is not the answer to farmers' problem. Instead the government should provide all growth inputs and incentive to the farmers like quality seeds at affordable rates, assured irrigation, remunerative minimum support price and enhanced subsidy, security from exploitation by middle men, assured market so that they are not driven to go for stress selling of their produces.

Source: The Financial Express; 12 July, 2006

Policy Issues

11

Questions of Agricultural Productivity

Prime Minister Man Mohan Singh and his economists have been harping on increasing agricultural productivity ever since this group came into positions of power. Their usual complaint is that agricultural productivity of farmers in India is a drag on the growth rate of Gross Domestic Product. In order to increase productivity they have proposed farm mechanization, corporate farming, and Genetically Engineered seeds [GE seeds or GMOs].

Slave-like, even the officials of the Ministry of Agriculture have kept silent; in fact their silence reminds me of the submissive ways of landless agriculture labourers who would do anything for the landowners including performing all sorts of menial tasks in the landowners' household. Similarly the scientists and bureaucrats of the Agricultural Ministry have chosen to remain silent and plant all the vile and crafty variety of lies to justify and support the Masters' Voice.

Have they compared the productivity of industrial with radical farming method? The average yield of rice per hectare worldwide is a little over four metric tonnes; American rice growers have achieved marginally better yield. A group of radical farmers of Darveshpura village of Nalanda district of Bihar have achieved 22 metric tonnes per hectare, improving upon the world record of nineteen tonnes set by China.

These farmers did not use industrial methods. They used the System of Rice Intensification. Nalanda district is one of the poorest in the country despite the fact that had world famous Nalanda University established in 6thC BC which is now being resurrected under the supervision of Nobel Laureate Dr. Amartya Sen. Has the Minister of Agriculture taken note of this fact? Fittingly, this year's award for excellence in agriculture has gone to Madhya Pradesh [First Prize] and Bihar [Second Prize] and they don't use GE seeds. The System of Rice Intensification is not based on GE or GM seeds or expensive fertilizers and pesticides. And this is just one example.

With proper institutional support, China with much less usable land could increase horticulture production from 60 million metric tonnes [MMT] in 1980 to cross 450MMT in 2003, over seven times increase, and now it is far ahead of India's 140-150 MMT. In every major crop including rice, China out performs India and that is largely due to careful introduction of R&D support to farmers. Where is India? Its horticulture production is barely crawling up the productivity ladder. Why? As far back as in 1920s Sir Howard, then director of Pusa Research Station, had warned that 'Indian scientists are behaving like bureaucrats.' The ICAR institutions today are known for mainly replicating work. In contrast, the Chinese have made their institutions result and reward oriented.

The argument that industrial agriculture is more efficient and other systems can't feed India is fallacious. The modern industrial agriculture system of the western nations is built upon cheap energy, mechanization, mono-cropping, large use of chemicals, and huge government subsidies; it is not only dependent upon scarce resources, it is also wasteful estimated at 50% of the ex-farm produce by weight as a recent report published in the UK suggests. A typical American farmer, even with 8,000 to 10,000 acre holding, gets about 5% gross return on revenue (RoR), barely able to meet interest charges. Therefore, replication of western model is unlikely to achieve food and nutrition security in India.

The second issue in productivity measurement is how wasteful is the system. Ministry of Food Processing Industries estimated that India wastes Rs. 50,000 crore worth of food and industrialization of food system will reduce that wastage. Other estimates suggest that food wastage is of the order of magnitude anything between 20 to 40% of gross food output. Another recent estimate based on Government data suggests that 21 MMT of food grains were spoiled because of unscientific storage in Government warehouses.

Is that a productivity problem? When this Government can't even identify that the problem lies with the scale of wastage after the food has been produced, it is essentially a problem of inadequate scientific storage, safe transportation, further storage at distribution or holding points and finally with retailers, what specific measures does it have in mind to increase productivity? If a farmer can produce 22 tonnes of rice per hectare why he be blamed for low productivity?

Has the Government done anything in the past fifty years to expand storage facilities? Has the Government done anything substantial to transfer radical agricultural technology to the farmers? What is the Agricultural Ministry doing? Why have they become shameless salesmen of Multinational Food and Seed Corporations? Will the children or grandchildren of these officers live a healthy life after these corporations take control of our food system? Think about that.

Source : Editorial; Kisan Ki Awaaz; Feb, 2013;[http://www.kisankiawaaz.org/Questions%20of%20Agricultural%20Productivity.html]

12

What Development?

The people of India are sick and tired of government's approach to economic and social development. Development does not mean GDP growth alone; there are other factors that contribute to production and productivity growth. Most important factor in long term growth is the quality of education system.

Instead of strengthening the school system and institutions of higher learning especially science and engineering, these have been weakened to a level where human resource quality is progressively declining in nearly all areas except those that are sponsored by multinational corporations. Even adult literacy programme, started in 1951 with much hope promised by Dr Zakir Hussain, failed; the actual number of illiterate adult is perhaps more than the population of India in 1947.

In the National Development Council meeting the Prime Minister said he wants to remove farmers from farming. Would the Prime Minister assure the nation how is he going to employ these 148 million farming households. There are already 93 million slum dwellers in urban India; does he want to add millions begging in the streets?

Where is the employment for the millions of farmers in the country. Did it ever occur to him that only farmers can grow food for the nation? During his neo-conservative policies over one quarter million farmers have already committed suicide but he remains unmoved. Agrarian distress has reached such extreme level that younger generation does not want to be part of it. Does the government want to add another million to the statistics of death by suicide?

It is worth pointing out that around five years ago, top executives of American seed and agriculture multinational corporations were in a meeting in the Department of Science and Technology and they told the Indian officers that they can do with just one million farmers. When the

BKS, Maharastra State wing felicitating Dr. Krishan Bir Chaudhary in national convention, Ahmadnagar (MHR) - 27.Dec.2012

Indian officers reminded them that 148 million households are engaged in farming, the MNC executives told them, as one insider said, "They are your problem." Is the government taking orders from American and European MNCs?

We do have a Ministry of Health and Family Welfare, which should be renamed Ministry of Sickness and Disease. Look carefully: the main focus of this ministry for the past two decades has been birth control and vaccination. Look at any document and these two issues predominate over all other issues of health. Their main focus today is to vaccinate women and children knowing fully well that vaccination induced debility are destroying the future of billions around the world.

Why is the Health Ministry pushing vaccines when doctors have warned that it is causing autism in children? Is the price worth destroying millions of children at genetic level? Improvement in hygiene and sanitation can alone prevent most of the bacterial diseases and improved nutrition alone can prevent virus induced infection.

This is not rocket science; these have been known since 1900 AD. Should big multinational pharmaceutical firms be allowed to play with the lives of millions of Indians?

Industrial development painstakingly built with public money since the 1950s are being slowly but inexorably privatized. It is again well known that strategic assets of a nation should not be privatized and these assets include ports, airports, railways, energy sources and energy infrastructure, seeds and infrastructure support to nation's agriculture and food security system.

Yet these are the very sectors which this government wants to sell off to multi- national corporations. Every system that has sustained us for over 8,000 years as a highly evolved civilization is today under threat. Yet government economists continue to parrot our development in terms of just one single statistics called GDP.

Source : Editorial; Kisan Ki Awaaz; Jan, 2013; [http://www.kisankiawaaz.org /What%20development.html]

13

UPA's Cash Transfer Policy

The mainstream media has hailed 'CASH TRANSFER' as a brilliant move by Dr. Man Mohan Singh and his economists. Is it a new type of Vote Gathering Machine? Stuff the poor with cash to win their votes.

Who gets the sops? What are the criteria?

How many landless and migrant labourers have identification papers or Aadhar card? How many of the truly indigent have a bank account? What is the basis of allocation? How many BPL card holders own land? Banking facilities are not available in large number of villages.

The age old proven dictum of 'train a man to catch fish instead of giving him a fish' does not enter the desiccated heads of the planners in the government. Why is this UPA Government adopting Non Productive Development Model when the nation needs Productive Growth Development Model both in industry and agriculture? Since past two decades irrigation facilities have not increased and the soil health is declining.

Water mismanagement is seriously hurting agricultural productivity and employment in small and cottage industries of the agro-processing sectors. There is no proper planning by the planners in these most crucial sectors. The present model of construction of airports, highways, shopping malls, and high rise apartments through the unholy nexus between real estate mafia and the bureaucrats serves the need of 15% Indians, Indian crony capitalists and MNC's. What about the 85% Asli Bharat? The share market, import & real estate are the bases of present GDP growth, not real growth and unemployment and under-employment is increasing in the country.

The same policies were adopted by Sub Saharan African Countries and now they have became perpetually dependent on food aid from rich countries. That actually is the agenda of United Nations Framework Team for India and South Asia. As many as 85% Indian farmers are small and marginal and they have less than two acres of land to grow food and yet they feed this nation of 1.2 billion souls. Majority do not have BPL card.

How will these farmers and farm hands get the fertilizer subsidy? The government said that only BPL card holders are entitled to benefit from direct cash transfer. Is the government covertly planning to eliminate fertilizers subsidy? Will it not increase the cost of agriculture production in the country? Will it not destroy the viability of small and marginal farmers' economic basis?

Is the government planning to stop the Public Distribution System? Would they in the future also stop the procurement of agriculture produce from farmers at minimum support price? Analysts say that this is precisely the agenda of the United Nations Framework Team which is covertly influencing India's economic policy to eventually destroy its agricultural base.

This Government proposes to transfer 3.2 lakh crore to 10 crore households which comes to Rs 3,200 rupees per household per month. Various surveys across the country strongly indicate that if a household fails to earn Rs 8,000 per month it can't insure three meals for the family, proper education, social obligations and medical care. Rupees eight thousand multiplied by twelve months in a Gregorian calendar adds up to rupees 96,000.

The Rs. 3200 per month or 38,400 per year of cash transfers to BPL households promised by the UPA-II falls short of actual cash needs of a typical rural household while destroying a complex food production system. Since majority of BPL households are landless labourers, who will get the cash for votes? The Sengupta, Saxena and Tendulkar committees, respectively, estimated 77%, 50% and 37% percent of our population as poor, respectively. That roughly works out to be 90, 60 and 45 crore Indians, respectively for each method of assessment, as one analyst says.

The Rs 3,200 transfer does not even match the so called economists' estimate that Rupees 28 per day per capita consumption expenditure means that person is not poor. Rs. 28 per capita per day comes to Rs 10,220 per person per annum; for a family of five, Rs. 51,100 per annum. UPA-II is fooling the nation's voters and this looks like a plan for slow death of the poor people of India.

The government wants to eliminate subsidies on PDS, fuel, fertilizer, and wage support programme like MGNREGA. A recent report says Rs. 528,000 crore was given to private corporations as subsidy. Has the UPA any scheme to reduce that subsidy? Actually the government has forgotten economics and is towing the line adopted by Western Governments to impoverish the 99% and enrich the 1%.

Source : Editorial; Kisan Ki Awaaz; Dec, 2012; [http://www.kisankiawaaz.org /UPA%E2%80%99s%20Cash%20transfer%20policy.html]

14

Note of Caution to Government to Protect Farmers from MNCs' Nuisance

The Government has opened the doors for foreign direct investment (FDI) in the multi-brand retail chain in the country, totally unmindful of the harm likely to be caused to the farmers. While pushing for what is called the "big ticket reforms", it ignored the voices of the opposition parties and some of its allies. The Government also turned a Nelson's Eye to the widespread resentment of farmers and consumer organizations across the country. The big ticket reforms were undertaken at the behest of powers outside the country - the mighty food MNCs and their patron, the United States of America. With the backing of these mighty forces, the Government conveniently ignored the opposition at home.

The government hopes that a lot of foreign capital will flow-in which would push the GDP growth to the magic 9% and above. FDIs and technology transfer are needed for the growth of the economy in select sectors which are capital and technology intensive like heavy industries, aviation, railways, shipping and others. FDIs are definitely not required in retail chains. The traders in this country have enough money and knowledge at their disposal to run the business. Handing over the retail business in the country to the mighty MNCs would lead to the exploitation of farmers and consumers. In whichever developed countries the trend of corporatization of retail shops began, they experienced distress of farmers and local communities. The UK Government had to appoint Competition Commission to undertake a study and the report showed awful distress to farmers but the Government was helpless as it had no elbow space to deal with the situation. In India, the Government while opening the doors to FDIs in multi-brand retail chains has made some cosmetic safeguards, but not enough to safeguard the interests of farmers. The Government should establish a National Authority to act as a watchdog with majority representatives from the farming community. This Authority needs to have the power and resources to directly intervene and take action against malpractices and misdoings of the retail chains without waiting for a written complaint.

All the MNCs and domestic corporate houses in the retail chain should reveal their sources of supply and buying prices. Maximum Retail Price (MRP) must not be left open for retailers and there needs to be a reasonable cap on MRP in proportion to the input cost. The retail chains should ensure minimum 60% share of retail price to producers of milk, fruits and vegetables. There is a need to regulate and monitor contact farming to protect farmers' interests and see that their land is not confiscated. Payment to farmers should be within a month of procurement and no unreasonable rejection of farmers produce in the name of quality and standard should be made.

The watchdog should ensure that no single retailer monopolizes procurement operations in an area. It should prohibit vertical agreements between retailers or intermediaries and seed and fertilizer companies. Food retailers or other agribusiness companies should not be allowed to corner and hoard food grains stocks under any circumstances. To prevent cornering of stocks by corporations, there should be rules for public disclosure of stock holding levels. Public agencies should be empowered to purchase food grains from the private holders at pre-specified prices.

Source : Editorial; Kisan Ki Awaaz; Nov, 2012;[http://www.kisankiawaaz.org /Note%20of%20Caution%20to%20Government%20to%20Protect%20Farmers%20fro m%20MNCs%20Nuisance.html]

15

ICAR digs for MNCs

Deputy Director General (DDG) of Indian Council of Agriculture Research (ICAR) Swapan K Datta wants a tie up with multinational seed companies to strengthen ICAR's marketing ability across the world. India's share in the 8.2 billion dollar global seeds' market is barely 1.2%. The oligopolistic global seed trade is controlled by the same cartel led by Monsanto that also control genetically engineered seeds.

ICAR is a publicly funded division of the Ministry of Agriculture. It's the second oldest research institution. Starting with Pusa Agriculture Institute in the late 19th C, it has grown into a national network of research institutions but its primary task is to develop seeds and farming methods for India's farmers.

Its most strategic and vital task is to ensure that sufficient quantity of quality seeds are available to farmers across India. Competing with global oligopoly is not its mandated task. It can't compete because it does not have the backing of a military power like that of the US and NATO.

Datta should note that Monsanto, Syngenta, Bayer, Dow, DuPont are part of the full spectrum domination strategy of US and NATO powers. Their seeds sell not because they are safe or of better quality or even of lower price; their seeds are rammed down the throat of the farmers. Datta, therefore, has been told to "tie-up" with these multinational corporations. But will India's seed reach the farmers ever and will Indian farmers and research institutions be compensated for conserving the unimaginable bio-diversity of seeds?

Monsanto's stated objective is to control global food supply through control over seeds. Can Datta justify this scandalous nexus with multinational seeds' corporations? These multinationals have not given the world any seed. Seed variety was developed by farmers.

Dr. Krishan Bir Chaudhary, inaugurating Bharatiya Krishak Samaj Convention, Bagalkot (Karnataka) on 3 may, 2012

The companies slowly took control over seed variety in North America and Europe; they want to control all seed varieties around the world. They need natural seeds to genetically engineer them for patenting. Once they engineer these seeds and file a patent claim, the seed ownership gets transferred to the companies.

What will Datta's ICAR get? Rs 5.75 per seed or Rs 5.75 for the whole lot that he wants these multinationals to market globally? The final sell off of Indian seeds to MNCs is being initiated by Swapan K Dutta. Here is the final nail in the coffin of Indian seeds.

One man can kill Indian Agriculture, India's food security and sovereignty. All those billions of rupees of investment by Indian tax-payers over past one hundred years is being handed over the same multinationals who are hell bent on destroying India as a nation.

This act itself is sufficient to open the eyes of Indian politicians to see how our research institutions are working on the directions of MNC's and how our agricultural scientists are dancing to the tune of MNC's.

Who has given the right to hand over more than four lakh plant species and farmers' seed varieties, the natural resources and the richest wealth of the country, to MONSANTO? Where has our national character gone? Under IPR regime if we hand-over our Gene Bank to MNC's then food security will be automatically under their control. In future food will be used as a weapon.

The GM technology is dangerous for human and animal health and will spoil the whole environment, many European countries opposing it. In future, MONSANTO wants to introduce the TERMINATOR gene (patented) in all the seeds to make them sterile. Why is the government playing with India's food security and people's lives?

Source : Editorial; Kisan Ki Awaaz; June, 2012;[http://www.kisankiawaaz.org/ICAR%20digs%20for%20MNCs.html]

16

Water Crisis and the neo liberal Privatization Agenda

If all the world's water could be contained in a one litre bottle, fresh water would represent about two drops; barely 2.5% of world's water is fresh water. Of this 70% is frozen in Antarctica and Greenland ice caps. About 30% is stored in lakes, ponds, underground aquifers, streams and rivers, easily and cheaply accessible. In all, barely one percent of total freshwater is available for human consumption.

The per capita availability of fresh water in India was estimated at 1,729 cubic meters per inhabitant per year during 2003-07, showing decline of 27% since 1983-87. In India agriculture is often blamed for consuming 84% of fresh water. Under the neo-liberal regime, the sectors that show higher GDP contribution are given more weight.

Since agriculture contributes far less, every policy maker is quick to blame our hapless farmers for drawing too much ground water and wasting it. It never occurs to them that today every river, including our Holy Ganges, is severely contaminated from untreated urban sewage and industrial effluents. The pressure on water resources comes from population growth and decline of ecosystems, particularly the Himalayan ecosystem. Deforestation of large areas in the Himalayas and the Deccan plateau has turned many perennial rivers seasonal and many seasonal rivers have dried up as witnessed over the past two decades across the Himalayan foothills.

The per capita daily drinking water need has been estimated at 3 litres per day [Government of India; Rajeev Gandhi Drinking Water Mission]. Applying this rule of the thumb, India would require 1.2 billion x 3 litres=3.6 billion litres of drinking water every day. Add to that water needs for cooking, washing, bathing and the total household requirement is assessed at 40 litres per capita per day which means 48 billion litres per day. Today 377 million Indians live in cities and that means that 15 billion litres of water is required every day for urban India.

Dr. Krishan Bir Chaudhary, addressing the seminar on Water Resources & Agriculture Productive Deenbandhu Chhotu Ram University of Science & Technology , Murthal Distt – Sonepat (India)

Instead of tackling the problem at ecosystem level and raising awareness for water conservation, this government wants to privatize water resources. This sort of response needs to be debated particularly in the light of the decision of the Honourable Supreme Court.

The Supreme Court in the matters of the WRIT PETITION (CIVIL) No. 423 of 2010 with WRIT PETITION (CIVIL) No. 10 of 2011 in the infamous spectrum allocation scam dated 2012, has defined 'natural resource' and clarified the role of the State.

It raised a fundamental question: 'Whether the Government has the right to alienate, transfer or distribute natural resources/national assets otherwise than by following a fair and transparent method consistent with the fundamentals of the equality clause enshrined in the Constitution?'

The detailed clarification, citing past judgments of various courts in similar matters and the Constitutional provisions are given in pages 68 to 75 and the main points are: "It must be noted that the constitutional mandate is that the natural resources belong to the people of this country "It is thus the duty of the government to provide complete protection to the natural resources as a trustee of the people at large the solemn duty of the State to protect the national interest and natural resources must always be used in the interest of the country and not private interests the State and/or its agencies/instrumentalities cannot give largesse to any person according to the sweet will and whims of the political entities and/or officers of the State Appearance of public justice is as important as doing justice. Nothing should be done which gives an appearance of bias, jobbery or nepotism."

No society has benefited from privatization of water. The manipulation of accessibility to this vital resource is so pernicious that in poor countries like Kenya water is more expensive than alcoholic drinks like beer.

Such moves will deny fundamental right of access to natural resource and that is against the principles of natural justice. Such moves should be summarily rejected.

Source : Editorial; Kisan Ki Awaaz; May, 2012;[http://www.kisankiawaaz.org /Water%20crisis%20and%20the%20neo-liberal%20privatization%20agenda.html]

17

MNC's will Capture Food Security through FDI in Retail

The rural India is facing agrarian crisis and 68% the country's population is dependent on agriculture. A many as 85% farmers are small and marginal with less than two acres of land. In India the retail sector gives employment to forty million people. We cannot compare ourselves with China because it is a collectivist society. India's priority is poor people's welfare and that is our primary national interest. When the Government cannot provide sufficient employment to the people then why is the government destroying the self-employed service sector of the economy. There are many examples in the developed countries that FDI does not benefit farmers.

For MNC's it's agri-business but for us it's a way of life. The FDI in Retail is nothing but a game-plan to control the food security of the country through the contact farming system. We will lose rights on our natural resources like land, water and the environment. Once the MNCs establish monopoly control over our natural resource base, the farmers will be finished exactly as the East India Company did and the people were starved and enslaved. Even in the USA people are opposing the American development model and policies of these corporations which known as the Occupy Wall Street Movement.

They always buy cheap and sell at the highest known price globally. FDI is the root cause of corruption in all developing countries. It is widely known that MNC's are known for lobbying by using money power to gain advantages and favors. Independent Farming and Trading are the pillars of strong India. Why is the government not ready to develop the infra-structure that supports these small self-employed people in the country.

Agriculture needs better marketing, better infrastructure and effective network in the country. The traditional retail culture cannot replace by the MNC's system. If the food security of the country comes to be controlled by these MNC's through there patented GM seeds and inputs in agriculture sector, the farmers and consumers both will be exploited as is happening in Europe and North America. The economy of the country will be controlled by them, because food is the main currency of India.

If FDI in multi-brand retail is allowed in India, then our farmers will suffer the same fate as their western counterparts. In the US, dairy farmers' share in consumer's dollar has declined from 52% in 1996 to 38% in 2009. Similarly, in the UK, the share of dairy farmers declined from 56% in 1996 to 38% in 2009. The wastage of fruits and vegetables due to poor storage in India is a false argument, because data clearly proves that wastage are much less in India as compared to every developed country developed country.

The more of a commodity, the large retailers buy in bulk at lower prices from agricultural producers. Studies have proved that Wal-Mart and other retail giants create a monopolistic market and the existing markets disappear. Wal-Mart's products come from China and has an annual turnover of 422 billion dollars in 15 countries with 2.1 million employees. Wal-Mart's procurement from China is the major source of its profits. The Government is not worried about Indian jobs and is more concerned about the interests of global retailers.

The argument that inflation will be reduced by reducing middlemen is false because they just replace small traders with a giant trader who collects higher profits. The Hyper markets displace diversity, quality and taste. The giant retail will not only destroy our retail, small manufaturers and farmers but also devastate our culture and the very social fabric. Studies of developed nations where global retailers have concentrated their business activities show that farmers have not benefitted anywhere. The high level and sophisticated standardization by giant retailers lead to immense wastage of farm produce.

The huge subsidies for farmers in developed nations show that FDI does not benefit farmers and as a result the Farmers are perishing over the globe. As far as the consumers are concerned they will be adversely affected too. Experience of countries like Mexico, Argentina, etc, show the supermarkets usually sell more expensive food than their small and informal outlets.

The protests in Arab Countries and other parts of the World amply reflect the destruction of small retail stores which has resulted in mass unemployment. Why global retailers are facing resistance in their own countries? The supporters are those people who have vested interests.

Source : Editorial; Kisan Ki Awaaz; April, 2012; [http://www.kisankiawaaz.org /MNC's%20Will%20Capture%20Food%20Security%20through%20FDI%20in%20Retai l.html]

18

Government should Implement the National Policy for Farmers

I said in my various farmers meetings and interaction with media during my Tamil Nadu farmers yatra (which started from Kanyakumari on 21st January and ended in Chennai on 26th January 2012) that the National Policy for Farmers, formulated in 2007, should be implemented. The farmers supported me, so did the scientific community.

This document defines a farmer as a person actively engaged in the economic and/or livelihood activity of growing crops and producing other primary agricultural commodities and will include all agricultural operational holders, cultivators, agricultural labourers, sharecroppers, tenants, poultry and livestock rearers, fisherpersons, beekeepers, gardeners, pastoralists, non-corporate planters and planting labourers, as well as persons engaged in various farming related occupations such as sericulture, vermi-culture, and agro-forestry.

The term will also include tribal families / persons engaged in shifting cultivation and in the collection, use and sale of minor and non-timber forest produce.

Special categories of farmers include three: tribal, pastoralists and others like urban farmers. For the first time we see urban farmers mentioned as a special category among plantation and Island farmers.

This is a step forward towards food and nutrition security of fast urbanizing India. Since 377 million people now live in its cities and towns, which is more than India's population in 1951, urban farming needs a boost at policy and technological level.

People must learn to grow food. Nearly every urban household grows some plants, mainly Tulsi, some flowers, or other decorative plants. Even in slums, people grow plants.

Dr. Krishan Bir Chaudhary launching Farmers Yatra from Kanyakumari, Tamilnadu on 21st January 2012, with Sh. Nallasamy, Sh. D. Gurusamy, President Tamilnadu BKS and other eminent farmer leaders.

If Cuba could produce 60% of its urban food requirements in its cities and towns, why can't we achieve at least 50%? Locally grown foods will be cheaper and more nutritious and a system can be created whereby the food reaches the kitchen without expending scarce fossil fuel energy.

We also need to feed the 93 million slum dwellers, many of whom are denied basic entitlements to food because many are migrants. The number of slum dwellers will rise to 110 million by 2017, a huge problem for urban food security.

And this will grow faster as agrarian crisis in rural areas worsens. At the moment the rural poor are starving from low cash income; the urban poor are also starving from low cash income.

The urban poor can be networked into the self-sufficient food systems on vacant lands, even the roof tops of their huts. They can grow sweet gourd [lauki, sitaphal], bitter gourd [karela] and feed themselves; the surplus can be sold to neighbours.

We will have to look into our urban planning processes. The next decade will see a closer coupling of rural and urban farming systems, issues of health and nutrition and closer inter-dependence of the rural and urban intellectual and scientific capital, resources and skills.

Therefore, the government should implement the National Policy for Farmers, it was placed in Parliament in November 2007, includes the following goal–"to introduce measures which can help to attract and retain youth in farming and processing of farm products for higher value addition, by making farming intellectually stimulating and economically rewarding."

This would pave the way for food and nutrition security of fast urbanizing India and also create a mechanism of transfer of intellectual capital to the urban poor and the rural farming communities. We need a symbiotic rural-urban system to ensure food and nutrition security and the National Policy can be tweaked to achieve that.

Source : Editorial; Kisan Ki Awaaz; Feb, 2012;[http://www.kisankiawaaz.org/Government%20should%20implement%20the%20National%20Policy%20for%20Farmers.html]

19

How Beneficial will be the Multinational Retail Chains

One vital question agitating the public mind is "Do we need foreign direct investment (FDI) in retail chain? If so, for what?" The Government says that FDI in retail chain will benefit the farmers and consumers and create about a crore of new jobs in the country. This tall claim of the Government is nothing but a hoax and without substance.

Usually the demand for allowing FDI is made by the sector or the industry which is in need of capital and technology. But in the case of allowing FDI in retail chain no such demand came from the farmers' or consumers' bodies. The fact is that Europe and the US are reeling under economic recession and the multinational companies based in these regions are eager to invest in the Third World countries to maximize profit.

The Indian Government in turn wants to increase the FDI inflows with the intention of arriving at the magic growth figure of 9% or 10%. If the Government is so ambitious of increasing the FDI inflows then why does it not encourage such FDI inflows in large industries and sectors which are capital starved and need modern technology? Globally the food business is controlled by few retail chains like Wal-Mart, Carrefour, Tesco, Ahold, Kreger, Rewe, Aldi, Ito-Yakado, Metro group, Intermarche.

Even a British Government report [Competition Commission, 2008] has said that the retail chains have squeezed out the farmers and put to heavy losses. The entry of retail chains has displaced thousands of independent shopkeepers. If this is the case of farmers and shopkeepers in the developed world what would be the fate of Indian farmers?

The entry of multinational retail chains in the country would endanger farmers' wellbeing, national food security and consumers' interests. The retail chains would control all the inputs including seeds and the farmer will be forced to grow what the retail chains want them to grow. Farmers have always been tied through iniquitous contracts to these multi-brand retailers.

But the produce of the farmers may be subjected to rejection on the basis of colour and size and on other fake norms. The retail chains would be free to import any food items from outside the country to run their business. Thus the multinational retail chains would control the food economy of the country, displacing farmers from their livelihood.

Globally Wal-Mart, Tesco, Carrefour and Metro group together provide around 3.1 million jobs. In this context, it is ridiculous for the government to say that retail chains in India would create 10 million [one crore] jobs in the country.

The real fact is that the retail chains would displace more than 40 million small retailers from their livelihood.

Finally the consumers will have to pay high prices for foods in malls. They will be deprived of sourcing cheap and fresh fruits and vegetables from local retailers. The rejected GM foods will be dumped in this country.

Source : Editorial; Kisan Ki Awaaz; Dec, 2011;[http://www.kisankiawaaz.org /How%20much%20beneficial%20will%20be%20the%20multinational%20retail%20ch ains%20for%20the%20country.html]

20

The Exercise of Framing Agricultural Policy

Much mid-night oil is being burned to finalize the agricultural component of the 12th Five Year Plan, and many states are working overtime to prepare a wish list to ensure three healthy meals to 1.2 billion hungry Indians. The economists are still not sure what should be the minimum cash income of BPL households, and majority of our small and marginal farmers fall in that category.

Their objective appears to be to ensure cash income to help them buy food and stave of hunger and starvation. It is ironic that the people who feed the nation should be the concern of economists and planners in terms of food. Therefore, each plan document is replete with a long wish list of productivity improvement, diversification into fishery, goatery, piggery, mushroom cultivation, silvi-culture, etc; you name anything that can be grown on a piece of land or the roof top of a Katcha house, they are in the plan document.

The multinational corporations and their local partners in seeds, food processing, distribution and retail have another priority. They want to modify agricultural plans and policies to extract as many sops and as much regulatory protection as they can manage; they want monopolistic control over the market by eliminating competition. That is the typical behaviour and conduct of transnational corporations. Neither the economists nor the corporations have the interest of farmers and consumers in mind and this is a problem India has suffered for far too long.

As per recent data and trends, nearly all rivers and majority of lakes and ponds are now contaminated with industrial effluents, agricultural chemicals or rural and urban untreated wastes. Rainfall pattern is changing from climate forcing too. Hence, there is problem of availability of irrigation water much more acute in Punjab, Haryana and Western UP as also in nearly all rain fed regions.

India is short of conventional energy resources yet the drive for mechanization of even small farms is being pushed. Diesel price increase directly affects farmers cost of production; non-availability can cause collapse of agriculture because farmers have come to heavily depend upon bought-in energy. Electricity supply in rural India is a big problem making farmers more dependent on diesel oil.

What is important for every state, particularly where majority of farmers are small and marginal with large landless households, is to plan for a system that ensures (a) food security, (b) nutrition security, (c) sufficient cash income from farming and related activities and (d) a healthy ecosystem which supports food production. It is time that planners recognize the centrality of ecosystems' role and within this broader framework the constraints of availability of land, water, energy and seeds.

Furthermore there is now a dire need for bottom-up planning, from habitation level to Gram Panchayats, Block and District levels. The role of the Central Government should be limited to supporting the required agricultural investments as desired by each state.

Source : Editorial; Kisan Ki Awaaz; Nov, 2011;[http://www.kisankiawaaz.org/The%20exercise%20of%20framing%20agricultural%20policy.html]

21

Policies Ruining the Agriculture Sector

The corporates have started their game-plan to capture the agriculture sector. Policy drift and absence of a truly national policy on agriculture have thrown rural economy into disarray and may prove even more disastrous if the situation is not handled with deftness. Agrarian well being has been thrown at the mercy of market driven forces. It is outrageous and totally unacceptable. Thinning out of the euphemistic smoke screen reveals the ugly face of the sinister design of the agri-business multinationals and corporate houses to cast their gripping hold on India's farmland.

Colossal and enormous loss that will be inflicted by contract or corporate farming on our traditional agro-system, social-cultural ethos, growth factors, biodiversity and the eco-system is beyond comprehension, making it absolutely difficult to reverse the situation. Sabotaging and dismantling a system that sustained the life and times of a vast majority of our polity and provided sustenance to the agro-rural segment of our society for centuries will usher India into a state of economic subjugation.

Alluring unsuspecting simple farmers by eulogizing corporate farming as the panacea for their socio-economic emancipation, the corporate sector will tighten their grip on India's farmland, turning the source of sustenance and livelihood for the peasantry into a market driven entity. It is highly erroneous to belittle its importance in India's context which predominantly is an agricultural country. It is capable of strengthening and sustaining all round economic growth of the country even in times of worst of Industrial recession and stagnation that may come-up on account of unforeseen fluctuations in international trade.

Agriculture in India still has enormous employment generation capability. The decline in rural economy can be arrested only if the policies are made genuinely realistic and crafted out on participative basis by associating the stake holders and in complete consonance to suit their needs and aspirations. What actually needed is a strategic planning. To put it plainly, we can say that destabilization of our rural economy is a clear case of failure to evolve a strategy compatible to our needs. Small and marginal farmers, having less than two acres of land holding and facing various natural and man-made constraints, still manage to sustain their livelihood from small family farms and feed the nation. It is highly erroneous to think that market driven commercial venture like corporate farming can ever be trusted as a dependable source of income and a rewarding venture. Such diabolical policies will further debase the agro-rural segment and will only serve the interest of the MNC's and the Indian corporate sector.

Growth and development of agriculture should be all about sustainability, self sufficiency and self-reliance. However, the policies being pursued defies all logic and reflect deep cynicism that despite abundant resources India has been forced into a food importing nation. Something seems to have seriously gone wrong. There is urgent need for the policy makers to peep into the causes of failures and re-fixing of priorities by bringing agriculture in the centre stage of economic growth and development, otherwise the way we are moving, nothing can check us from becoming a satellite nation.

Source : Editorial; Kisan Ki Awaaz; Oct, 2011;[http://www.kisankiawaaz.org/ Policies%20ruining%20the%20agriculture%20sector.html]

22

Why is FICCI asking for 50% Land to be Acquired?

Why is FICCI asking for 50% land of small and marginal farmers to be aquired by the government for the corporate for their real - estate and shopping malls' business? Already government has given tax exemptions to all the corporates in the name of SEZs.

It should be noted that there is no such example anywhere as India's SEZ policy. The SEZs in China are owned by the government and essentially developed by the government on non-farming land. On the other hand the the Indian SEZ policy provides for development of these zones in the government, private or joint sector.

Not only that, the complete bouquet of exemptions is astounding as shown below.

Except for activities included in the negative list, 100 per cent FDI is permitted for all investments in SEZ. SEZ units are required to be net foreign-exchange earners and are not subject to any minimum value addition norms or export obligations. Facilities in the SEZ may retain 100 per cent foreign-exchange receipts in that currency. Goods flowing into SEZ from a domestic tariff area (DTA) are treated as exports, while goods coming from the SEZ into a DTA are treated as imports. SEZ developers also enjoy a 10-year "tax holiday".

The size of an SEZ varies depending on the nature of the SEZ. At least 50 per cent of the area of multi-product or sector-specific SEZ must be used for export purposes. The rest can include malls, hotels, educational institutions, etc. 100 per cent FDI is permitted for SEZ franchisees in providing basic telephone services in SEZs.

SEZs enjoy complete exemption from Customs and Central Excise duties on import and procurement of capital goods, raw materials, consumables, spares, etc. Exemption from Central Sales Tax and Service Tax. No import license is required. They are also exempted from industrial licensing for items reserved for the small-scale sector. No routine examination by Customs for export and import cargo is required. SEZ units are exempted from Income tax for a period of 10 years and also exempted from Service Tax.

The report of the Comptroller and Auditor General [CAG] has estimated that SEZ has directly caused revenue loss of around Rs 1,74,000 crore.

It should be noted that whilst the Government had assured the people that 5 million jobs will be created, in actual fact only about 700,000 jobs have been created. These SEZs have further caused livelihood losses of 20 million people in India.

With all these benefits and exemptions and non-performing SEZs, why does FICCI still wants more sops from the government?

Why do the corporates want to forcibly grab prime agricultural land of farmers? Agriculture is not a business; it's a way of life for 70% of the country's population.

Source : Editorial; Kisan Ki Awaaz; Sept, 2011;[http://www.kisankiawaaz.org/How%20the%20FICCI%20is%20asking%20for%2050%20land%20to%20be%20Aquired.html]

23

Price Control Mechanism
has to be Relooked

There is a vast mechanism in the country to check any volatility or abnormal rise in prices of essential commodities. Despite expending billions of rupees to run the price control machinery at the Central and State levels, this mechanism has been rendered ineffective and counterproductive. The present surge in prices of essential commodities and vegetables of common use has gone out of hand. The government machinery has lost its effectiveness to bring the situation under control.

The Government has made an arrangement to monitor daily and weekly prices of 17 essential commodities across the country. Its main purpose is to check demand and supply situation and find balance if there is any supply problem in the country. The government undertakes short term measures like import and export to control and augment supplies so that common man is cushioned from price rise.

There is a special cell in the economic division of finance ministry for this purpose. A group of ministers has been setup to take necessary action on price situation and to suggest remedial measures. The group recently took decision to release additional quotas of wheat and rice to states and extent ban on export of pulses. It is worth mentioning that on the recommendation of this Group of Ministers the Government of India on 7th Sept., 2010 had allotted 25 lakh tonnes extra food grains to states. Off take by the state governments from this allotted quantity was only six percent by 9th Nov., 2010. This implies that states have enough food grain for BPL families.

A committee of secretaries under the chairmanship of the cabinet secretary, the strongest committee on prices, is also there to save common man from abnormal price rise. Many times it has been noticed that the government machinery wakes up only when the situation has gone out of control. Sugar market's volatility and recent surge in onion prices are fresh examples of lack of efficiency of the present mechanism for price monitoring and control.

The issue of price rise was discussed nine times in the Parliament since 2004, three times in 2006. The Reserve Bank of India on its part has undertaken steps to control money supply to curve inflationary pressure. The Central Government advises the State Governments to take effective action to check hoarding and black marketing under Essential Commodities Act 1955 and Prevention of Black Marketing Act 1988.

There is always a crisis on the price front. Sometimes commodity 'A' troubles the common man, at other times commodity 'B' or 'C' . People are losing faith in the system. It is visible that neither short term, nor long term measures are working to control the price rise. Unscrupulous hoarders and profiteers are exploiting the situation. Frequently hoarders create artificial crisis in the market. Common man and poor farm producers are real looser.

The situation will improve only if government makes effective use of its power to reduce the gap between the prices at which farmers sell their produces and the prices at which the consumer buy the products at retail shops. If the government has will-power, it can surely find the way.

Source : Editorial; Kisan Ki Awaaz; Jan, 2011:[http://www.kisankiawaaz.org/ Price%20Control%20mechanism%20has%20to%20bee%20relooked.html]

24

Farmers hold up more than half
the sky, but Losing Land

Actually in the white hot heat of "development" politicians and people have forgotten the farmers who produce the food that fuels 8% GDP growth. In January 2008, at Davos, international summit, the votaries of 'globalization' were fawning over their achievements inside the plush hotel, while outside the common folks, supposed to be the beneficiaries were worried about food, water and fuel...all coming under control of global criminal corporations.

Indian farmers were not there or they would have added land. The way both Indian and foreign multinationals are forcibly stealing land from them would pauperize not one or two but millions of farmers in India. India has about 148 million farming households eking out a living of farming activities. This is about 65% of total households in India. Insiders say that in a meeting of agri-biz multinational corporations, one of the executives said, "we just need one million farmers."

To this the officials of the Indian Government asked, "What we would do with the remaining?" The answer was: "They are the Government's problem!" Lest you forget: the corporations' agenda is to take control of the natural resources and enslave the farmers . It is the natural resources that sustains us but the entire edifice of the current development agenda is to control land, water and diversity of our seeds. None of these were created by any corporation. Those studying the impending energy crisis have come to a conclusion that when the world comes to the end of the finite fossil fuel era, the United States of America alone would require 50 million farmers to feed Americans.

The cheap oil and gas era is likely to come to an end by 2015 and it really does not matter when it comes, today, tomorrow, in 2015 or 2030. Cheap energy allowed industrialized farming, corporate control of food production and high degree of labour efficiency: far less human muscle was required to produce food as compared to the period 70-80 years ago.

When cheap energy will be unavailable who will feed our bureaucrats, army and the rest of us? So, the prescription by world's top agri-biz corporations that they need only one million farmers is not only misleading but a ruse to push India into a situation of food crisis, engineered food shortages, mass hunger, starvation and deaths.

In the era of energy shortage if America would need 50 million farmers, India would need 150 million. Make no mistake: we need a policy for restoration of the peasantries. But can we do it? It took centuries for Indian farmers to make India world's largest producer of food and fibre; in less than three decades they have lost all their skills and committing suicide on a scale never seen in this country.

The 11th Plan document is ambiguous. It talks of supporting small farmers "but" also medium and big farmers! That "But" speaks of policy ambiguity of this regime.

We think that policy-makers and politicians may come and go but certain policy postulates need to be absolutely clear on the ownership and management of agricultural land. That is the most important issue right now. Land Acquisition Act [LA Act] of 1894 vintage still remains on our statute books.

The SEZ Act has further eroded the natural rights of the peasantries. Since 1947, forests have declined, pastures have declined by at least 20,000 hectare, and ecosystems have declined to the extent that almost all perennial rivers are stressed and seasonal rivers have dried up.

It is time that further disruption of peasantries from their lands are stopped and they are given all the help required to grow enough food for India's 1.18 billion and growing. Six crore farmers have been displaced since independence and more then two Lack farmers have committed suicides.

No farmland should be acquired in future and this is not a fad of environmentalists or supporters of farmers; this is a compulsion now. Enough is enough. The situation today is identical to the one that Mahatma Gandhi saw with his own eyes in Champaran.

Source : Editorial; Kisan Ki Awaaz; Sept, 2010;[http://www.kisankiawaaz.org/Farmers%20hold%20up%20more%20than%20half%20the%20sky,%20but%20losing%20land.html]

25

The National Food Security Act

The Food Security is a state that protects and fulfills the "Right to Food". The Concept of Right to Food is the embodiment of economical, cultural and nutritional values. The National Advisory Council (NAC) constituted at the behest of the UPA chairperson, Mrs. Sonia Gandhi is trying to reach at least the bottom of these values while implementing the proposed National Food Security Act.

With this mind set, the NAC decided in its recent meeting to redraft the present Bill. As such the Bill is not going to be presented in the current session of the Parliament. The politics of food knows no such values. For the people, "Food" is a human right to be achieved to protect their right to live guaranteed under Article 21 of the Constitution. For politicians, "Food" is a political tool to make use of the peoples' right to vote as a democratic exercise.

As such the political will of our body politics is lacking the potentials of enacting a "Law on Food Security". An in depth reading of the National Food Security Bill in its present form and the decisions of the recent NAC meeting is indicating that the universal acceptance of a food security law including all Blocs either geographical or social seems a distant feature. As of now, the people have to be content with the present food entitlement programmes in poverty reduction.

The mandate of the proposed Act to establish of Food security Fund at the Union Level and State Levels to compensate in case of failure to food provision is itself a nullifying the justifiability of the right to food. With this compensatory provision, failure to provide guaranteed food will allow the victims to receive the dole but not the required food as a fundamental and basic right. The quantitative restriction of 25 kg or 35 Kg per family per month food supply without guaranteeing the quality of the contents is another flaw.

The exclusion of Universal Public Distribution System (UPDS) and the inclusion of Targeted Public Distribution System (TPDS) is another anomaly. The NAC tries to understand the reality of universalisation but is unable to match it with the economic logic.   The NAC with its political, social and humanitarian action oriented blend of leadership will find it hard to fulfil the process of ensuring food security to our millions. Effective implementation of agrarian reforms with economic pragmatism is required. Unless the policies address the livelihood issues of small and marginal farmers and landless agricultural workers.

In this constructive and inclusive development agenda of providing food security to the suffering mass there should be no scope for party politicians in the name of framing a "Law For Food Security".   In this endeavor, the Ruling Governments by adopting the "Doctrine of Trusteeship over Natural Resources" the NAC may consider its domain to expand the opportunity of making the

Right to Food Law for the country and the people who are hungry, malnourished and literally innocent even to understand the root causes of the abject poverty stricken to them. The coalition partners of the present UPA government wish to demonstrate their concern for the poor, marginalized and neglected communities with this right to food and prove that the governance of the country is for the people, by the people and of the people. This is completely inconvincing and misleading.

Source : Editorial; Kisan Ki Awaaz; Aug, 2010;[http://www.kisankiawaaz.org/The%20National%20Food%20Security%20Act.html]

26

The Racketeering in Warehousing and Storage Facility

Warehousing plays a crucial role in the food value chain, from production to distribution and retailing. Food is produced by farmers. They produce food, feed, fodder and fibre that sustain this nation of 1.24 billion people. They need warehousing facility and all those support systems promised to them six decades ago by Nehru's government.

Instead, the UPA-II wants to establish COLD-CHAIN that only supports the corporations who buy raw foods cheap from dying farmers, convert them into processed foods for domestic and export consumption. The farmers have been short changed time and time again.

Please recall that the Reserve Bank of India *RBI+ in 1946 formed an "India Rural Credit Survey Committee" that recommended the establishment of warehouses to strengthen the rural credit and marketing. As a result of the recommendations of the Committee, the Government of India enacted the Agricultural Produce (Development and Warehousing) Corporation Act, 1956. Among the four key recommendations were:

Scientific storage: In Warehouses the stored produce could be protected from the vagaries of weather and pests [rodents, insects, etc.] and substantially reduce post harvest losses that according to various estimates ranges between Rs. 20,000 to 50,000 annually.

Financing: Warehouses were mandated to meet the financial needs of farmers with a provision for issuing 'WAREHOUSING RECEIPT.' It was felt that the farmer could obtain cash support from any bank against the receipt for his/her stock.

Price stabilization: Warehouses would help in regulating the price levels by regulating the supply of food grain in the markets. More goods from the buffer would be released when supplies are less. They would also advice farmers when to sell when there was glut in the market. Thus they would monitor the supply and demand in the interest of farmers. It would also prevent distress sale.

Extension services: Provision for appointing technical officers was made at each warehouse who would advice farmers on seeds, fertigation, irrigation, etc, for various crops consistent with data from the market.

Strategies to reach out to the farmers were further refined with talks of smaller rural warehouses to large nodal warehouses. It should be noted during the early days, the CWCs and SWCs had highly committed staff. They performed admirably and that was responsible in no small measure for rapid agricultural growth during late 50s right up to early seventies.

Should Government promote cold chain when they could not provide vital warehousing services to 148 million farming households? Who would benefit from cold chain? Farmers or multi-national corporations? Why should investments in cold chain be subsidized as it is planned? If people came to know of the true purpose of Warehousing Act of 1956, taxpayers wouldn't mind subsidizing hundreds of thousands of warehouses [large, medium, small or micro].

However, if people get to the truth of how warehousing corporations short changed farmers and how cold chains are going to indirectly subsidize large food corporations and many multinational food and agri-business corporations, I doubt even one person would support this sort of lop-sided policy.

The governments [both Central and State] need to re-focus their effort at rapidly expanding warehousing capacity for farmers, stop indirectly subsidizing large firms' control over small producers through cold chains.

Most importantly, the government must revive the original concept enshrined in the Warehousing Act of 1956 because the situation for farmers is deteriorating by the day. It will not be too long before India faces a major food crisis.

Source : Editorial; Kisan Ki Awaaz; June, 2010;[http://www.kisankiawaaz.org/Farmers%20shortchanged%20again%20The%20racketeering%20in%20warehousing%20and%20storage%20facility.html]

27

ICAR should put on the Right Track

Agricultural Research in India is fast moving towards a blind alley. Indian Council of Agricultural Research (ICAR), the country's highest research policy and planning body has become a liability to the public exchequer. It has no idea, understanding and appreciation of research priorities for a predominantly agricultural country like India. What actually ails the functioning of ICAR is the inherent weakness of its leadership, propensity to bend under foreign pressures and inefficient management.

And now, to make the matter worse, there are findings of Swaminathan and Mashelkar Commission that are irrelevant as they do not fit into our agro-system. Their findings have overlooked the needs of Indian agriculture and do not care to identify the real malady and suggest constructive and realistic remedial measures. The implementation of the findings of Swaminathan and Mashelkar Commission report will make our agro-system captive of MNCs for all time.

The same bunch of advisors who were responsible for misguiding and misdirecting the entire system of policy planning and identifying priorities vis-a-vis the agricultural sector and agricultural research for a long time can again be held responsible for manoeuvring with the UPA Government solely to speed up promotion and expansion of Biotechnology in the country.

Their advocacy for launching a Second Green Revolution is deceitful as it is nothing but a conspiratorial ploy of making way for GMOs whereas elsewhere in the world including even developed countries, these technologies are facing stiff public resistance on bio-ethical grounds. In sharp contrast, what the country actually needs is not biotechnology or nano-technology but firm policy direction and focussed research on biological diversity based ecological agriculture. There is now a strong international support for such natural resource based farming methods.

Enigmatical as ICAR has remained all these years right from its inception, one sure impression is that it has miserably failed to respond to the needs of the peasantry. In an era of stiff global competition where survival itself is at stake, the country can no longer afford the expensive and extravagant show piece like ICAR, totally inept and nepotistic, where agricultural research is manipulated and manoeuvred under the tutelage of globalist's interest.

Heading the ICAR is a matter of competence and commitment and having innovative ideas as per the need of the farmer-centric model based on natural resource management. Scientists must be fully committed towards prospective growth and development of agriculture.

Worst of all is the shabby treatment given by the UPA Government to the farmers' leaders and organizations by keeping them out of the decision making process. On the other hand, the same UPA policy makers have included lobbyists who are actually influencing major policies on food and agriculture.

The farmers' organizations and farmers' leaders must now be included in policy, planning and research as an absolute necessity if we are to ensure that India navigates smoothly and effectively through global competition.

Source : Editorial; Kisan Ki Awaaz; Jan, 2010;[http://www.kisankiawaaz.org/ I.C.A.R.%20should%20put%20on%20the%20right%20track.html]

28

GEAC should address
Health and Environmental Concerns

The government must clarify why it is setting up the National Biotechnology Regulatory Authority (NBRA), replacing the existing regulator Genetic Engineering Approval Committee (GEAC) which is already acting as a single window clearance for biotech products. If the government feels that the GEAC is incompetent and inefficient, it should bring it to the public knowledge.

The Supreme Court, in the course of hearing a writ petition seeking a moratorium on GM crops, had ordered some improvements for introducing transparency in the functioning of GEAC. The government had always defended the functioning of GEAC in the Supreme Court. Has it got any moral right now to say that GEAC is not functioning well and needs to be replaced by NBRA?

The fact is that the GEAC, without caring for any biosafety norms and transparency, has been very fast in the approval of GM crops with a view to benefit the multinational seed companies. Since 2002, GEAC approved over 175 Bt cotton hybrids, five events and one Bt cotton variety. It has conducted field trials of Bt. Brinjal, Bt. Okra, GM Mustard, Bt. Cabbage, GM Tomato, GM Groundnut and GM Potato.

The functioning of GEAC has been questioned by many independent scientists, like the founder director of the Centre for Cellular and Molecular Biology (CCMB), Pushpa Mittra Bhargava. He called for a total review of India's experience with Bt cotton, including how Bt technology was brought into the country.

He has also sought a two to three years moratorium on GM crops, unless and until proper independent studies are done on biosafety like pollen flow, seed germination, soil microbial activity, toxicity, allergenicity, DNA finger printing, proteomics analysis, and reproductive interferences. At the global level, independent scientists like Arpad Pusztai have questioned the safety of GM food.

Pusztai pointed out that "Well-designed studies, though few in number, show potentially worrisome biological effects of GM food, which the regulators have largely ignored. In India, there were reports of sheep mortality on account of grazing over Bt cotton fields in Andhra Pradesh, which the GEAC did not consider with seriousness.

There are reported cases of illegal imports of hazardous GM food, which are not approved in the country and the government has remained a mute spectator. Illegal imports of GM food are in violation of the Rules, 1989 of the Environment Protection Act, 1986.

The annual amendments to the Foreign Trade Policy made in April 2006 said unlabelled GM food import would attract penal action under Foreign Trade (Development and Regulation) Act, 1992. But this is not implemented in absence of guidelines. The panel of experts and stakeholders headed by the additional director-general of National Institute of Communicable Diseases, Shiv Lal had recommended mandatory labelling of GM food, irrespective of the threshold level.

But the recommendations were not implemented either by the health ministry or GEAC. Rather, the GEAC allowed free imports of oil extracted from GM soybeans without any labeling, tests and restrictions. The plan to set up NBRA is largely based on the recommendations of the two panels headed by MS Swaminathan and RA Mashelkar.

The suggestions made and apprehensions raised by the Indian Council of Medical Research (ICMR) in its paper "Regulatory Regime for Genetically Modified Foods : The Way Ahead " have not been considered.

Monsanto is charging a high technology fee, which has raised the prices of Bt. Cotton seeds and the issue is subjudice before the MRTP Act. There are fears that pollen flow from GM crops to non-GM crops may cause problems for farmers, who may be asked to pay high technology fee for their own seeds as had been the case with the Canadian farmer Percy Schmeiser and others.

Indian farmers, in many areas have suffered heavy losses on account of failure of Bt. Cotton. States like Kerala and Uttarakhand have banned GM crops and the Centre, through the NBRA, is planning to override states governments' power to regulate agriculture.

The government should make GEAC more accountable to address health and environmental concerns, rather than set up NBRA.

Source : The Financial Express; July 14, 2008; [http://www.financialexpress.com/news /GEAC-should-address-%20health-and-environmental-%20concerns/335290/0]

29

Futures Trade in Farm Commodities Wrangle Continues

The reluctance of a high-level committee set up by the Indian government to spell out whether futures trading in agricultural commodities has resulted in an escalation of food prices has revealed deep division over the issue.

Whatever conclusion the committee finally arrives at, farmers' bodies and food security specialists maintain that unchecked trading and speculation have played a pivotal role in the spiralling prices of food staples in India.

It is time the government rejected this neo-conservative and corporate-led agriculture model and replaced it with a farmer-centric one, leading famers' rights activist Krishan Bir Chaudhary told IPS.

Reacting to the interim report, Chaudhary said that the government was waffling over the serious issue of food prices and that the Sen Committee was skirting the issue of India's poorly regulated markets allowing traders and corporate entities to reap massive profits through manipulation in trading.

The government is aware what the real problem is. Futures trading in farm commodities under Indian conditions makes prices highly volatile, Chaudhary, a senior member of the ruling Congress party, said. Besides online trading, unofficial (dabba) trading - in which speculators are not required to maintain margins for trading in commodity futures and also do not have to pay transaction fees or taxes - is popular at India's exchanges.

The government's Forwards Marketing Commission has been unable to regulate this trading. Under these circumstances, what the government needs to do is ban futures trading in all agricultural commodities, Chaudhary stressed, because, in any case, it was not the farmers who were the beneficiaries but traders and corporate houses "that have been licensed to hoard and manipulate prices."

Agents of the corporate houses are even now scouring the rural areas, offering to buy up farm produce at prices far lower than what are being quoted on the futures exchanges or in the spot market, or at the newly created retail chains, Chaudhary said.

Although India achieved record grain production of 219.32 tonnes in 2007- 08, including 94.08 MMTs of rice, 74.81 MMT of wheat, 36.09 MMTs of coarse cereals, and 14.34 MMT of pulses - prices have soared uncontrollably and placed them out of the reach of vast sections of the population.

"Yet, the government is not critical about the neo-conservative architecture of the new economy that it has imposed upon the nation," Chaudhary said, "Now it is taking a piecemeal approach, banning exports and liberalising imports of selected commodities.

The government's annual Economic Survey 2007-08, released ahead of the budgetary exercise clearly says: "Direct participation of farmers in the commodity futures market is somewhat difficult at this stage as the large lot size, daily margining and high membership fees work as a deterrent to farmers' participation in these markets."

Chaudhary said the government was reluctant to ban futures trading in farm commodities because of a powerful lobby of traders that had reaped millions of dollars through speculative online trading in agricultural futures ever since this was allowed four years ago.

Turnover figures at the commodity exchanges for the 2006-07 financial year alone touched 926 million dollars. "This is a system that puts the lives of a vast chunk of the population at the mercy of a handful of speculators," said Chaudhary. "But they have tasted blood and are not going to let go so easily."

Source : IPS News; April 24, 2008;[http://www.ipsnews.net/2008/04/india-futures-trade-in-farm-commodities-wrangle-continues/]

30

Futures Caused the Market Manipulation

Futures trading in wheat, rice and pulses like tur and urad has been suspended by the Forward Markets Commission as it caused market manipulation, leading to a rise in prices. But, still, futures trading is being carried out in a number of agricultural commodities.

The government knows for certain that futures trading in farm commodities is the cause for market manipulation. Finance minister P Chidambaram, while presenting Budget 2008-09, slapped a commodities transaction tax (CTT) on options and futures on the lines of the existing securities transaction tax.

"The commodity futures have come of age in the country and should be treated at par with the equity market," he had said. Chidambaram also brought the commodity futures exchanges in the ambit of service tax. These measures were aimed at curbing manipulation.

However, the government's move is only a piecemeal approach although it has realised the damage futures trading in agricultural commodities can cause. It should nip the problem in the bud by banning futures trading in all agricultural products.

There is a wrong notion that the farmers are benefiting from the existing futures trading in the country. The farmers get the lowest price for their produce in the season at harvest and, thereafter, the produce passes into the hands of traders and corporate houses that manipulate high prices for commodities in the futures markets.

Farmers have no opportunity to participate in this. The Economic Survey 2007-08 clearly says: "Direct participation of farmers in the commodity futures market is somewhat difficult at this stage as the large lot size, daily margining and high membership fees ... work as a deterrent to farmers' participation in these markets. Farmers can directly benefit from the futures market if institutions are allowed to act as aggregators on behalf of the farmers."

Farmers have no time to participate directly in the futures markets. They have to prepare the field after harvest for the next crop. The concept that institutions or corporate houses should act as aggregators on behalf of farmers amounts to leaving the peasants at the mercy of these marketing giants.

The government has now gone into a panic mode as inflation, as measured by the point-to-point movement of the wholesale price index, reached a 40-month high at 7% for the week ended March 22, 2008. Yet, it is not totally critical about the neo-liberal architecture of the economy that it has imposed upon the nation.

It is taking a piecemeal approach like banning exports and liberalising imports of certain commodities. It is time the government rejected this neo-liberal and corporate-led agriculture model and replaced it by a farmer-centric one. There is no shortage of food either at the global or at the domestic level. According to a recent report of the International Grain Council (IGC), the world wheat production would be at 646 million tonnes (MT), an increase of 42 MT over the previous year, due to a 2.5% increase in the area under cultivation.

The global prices of maize were around $240 a tonne by March 27. The IGC forecasts global maize output to decline by 20 MT to 748 MT. Barley output would increase 10% to 148 MT. According to the official estimate, India has achieved record grain production of 219.32 MT in 2007-08, including 94.08 MT of rice, 74.81 MT of wheat, 36.09 MT of coarse cereals, and 14.34 MT of pulses.

The cotton output is estimated at 23.38 million bales of 170 kg each, an all-time record. The oilseeds output is estimated at 27.16 MT. Despite the good production, there is a deliberate manipulation of food prices both at the global and at the domestic levels.

At the global level, there are a few corporate players in the food business that buy produce from farmers cheap, hoard the stock and manipulate the prices. The bio-fuel programme in Europe and the US is also a contributing factor to price rise.

In India, too, the corporate houses and retail chains have been allowed to buy produce from farmers, hoard and manipulate the market.

The farmers do not gain in the process as they are paid relatively lower prices than what the corporate houses quote on the futures exchanges or in the spot market, or at what the retail chains sell to the consumers.

Source : The Financial Express; April 14, 2008;[http://www.financialexpress.com/news/Futures-caused-the-market-manipulation/296336/0]

31

Decontrol will Benefit Sugar Industry Not Farmers

Before finalising any decision on decontrol or deregulation of the sugar sector, policymakers should ensure that the desired benefits flow directly to farmers. Cane growers are pivots of the sugar sector, which cannot grow and progress if their concerns are not properly addressed. Industry is urging the government to decontrol the sugar sector so that they can reap more profits. It is a common belief among economists and bureaucrats that if the industry prospers, the benefits ultimately trickle down to the stakeholders. But unfortunately, past experience shows that this has not always been the case. The pricing of sugarcane still remains a contentious issue. The central government announces a statutory minimum price (SMP) for cane based on the recommendations of the Commission for Agricultural Costs & Prices.

The SMPs are artificially fixed low, and do not allow farmers to even recover their cost of production. The present UPA government has done farmers great injustice by increasing the cane basic recovery rate from 8.5% to 9%. But, as there aren't many high-yielding varieties of cane available in the country, the increase in basic recovery rate will not benefit many growers. Therefore, the basic recovery rate should be reduced from 9% to 8.5%. As SMPs are low, many state governments fix their own state advised prices to help farmers.

The state governments have a right to step in, as agriculture is a state subject under the Constitution. Some criticise the decision of these state governments as politically motivated. In this context, I would like to ask: what do we do when the assessments of economists fail to recognise reality? What is wrong if the political economy can solve the problem in the right way? The industry, in most cases, fails to make payments to cane growers within two weeks as required by law and the government remains a mute spectator. If the industry wants to make a deferred payment, they should pay the farmers with accumulated interest. I was chairman of the Indian Sugarcane Development Council under the Union agriculture ministry from November 1991 to February 1995. We deliberated a number of issues confronting the sugar sector.

We suggested that the total profits of the industry generated not only from sugar, but also from by-products like jaggery, molasses, ethanol, press mud and power cogeneration should be shared with cane growers. Unfortunately, the government's cane pricing policy does not take into consideration profits from these by-products. The industry says that the cane price should be fixed on the basis of a percentage of sugar sales realisations. This is absolutely ridiculous! Yes, I am for a common cane pricing policy for the country.

Cane prices should not only be remunerative but should also be fixed in such a way to ensure sharing of benefits from the sale of sugar and all by-products. In Uttar Pradesh, sugar mills collect cane from farmers through cooperatives. The cane cooperatives cause problems in payments to farmers. We suggested computerisation of cane stock received and payments made to make the process transparent. We also detected fraud in weighing centres for cane and suggested its remedy. We also suggested that the sugar development fund not only be deployed to rescue the industry, but also used for farmers. The fund should be used to develop new varieties of high-yielding cane and to promote nurseries. The fund can be use to clear dues owed to farmers, particularly when the industry fails to make payments in time. The industry can later replenish the amount to the fund with interest.

Source : The Financial Express; Jun 23, 2008;

32

For 750 Million... Humans

The unilateral opening up of retail holds few advantages for the country

The government has so far been zealous in its commitment to opening up the retail sector. And that too, much before the WTO tells us to do so. Big corporate houses and MNC's have been invited to invest in retail purportedly with the intention of attracting foreign direct investment (FDI) but mainly to benefit some mighty economic entities. And all this is being done without a care for the livelihood of 750 million farmers and over 40 million small retailers.

The negotiations on the General Agreement on Trade and Services (GATT) are still on so there is no call for any country to open up its retail sector just yet. However, taking a cue from a suggestion made in the Economic Survey 2004-05, policymakers are busy drafting a proposal to open up the retail sector for FDI. This, when even the mandated agenda of the government, the National Common Minimum Programme, does not call for its opening up.

On a global scale, a few retail chains dominate the retail business, the major ones being Wal-Mart, Home Depot and Kroger of the US, Carrefour of France and Royal Ahold of the Netherlands. Naturally, India is the next big market and these chains have been asserting to their governments that they demand in gats that this sector be opened up. This is a matter of great concern, for there's the possibility of local, small retail shops closing shop and creating unemployment. Many unemployed youth now survive by opening small retail businesses and they cannot compete with global retail majors who come backed with huge investments.

Of course, there are reports saying the supermarket majors can replace this loss of self-employment with additional employment generation. But it's a distant possibility. Policymakers and neo-liberal economists argue that farmers would get a better price for their produce with the entry of big retail chains and the elimination of middlemen.

But global experience has shown us that big retail chains have in the end formed cartels, curbed competition, caused misery to farmers and dictated high prices to consumers. The neo-liberals should know that competition is fair play in an economy and not the play of a few mighty entities. The British Retail Planning Forum (embarrassingly, financed by the supermarkets themselves) discovered in 1998 that every time a large supermarket opens, on an average 276 jobs are lost.

The UK Competition Commission report of 2001 has also expressed concern over the loss of jobs. The New Economic Foundation found that £10 spent on a local organic box scheme in Cornwall generates £25 for the local economy within a radius of 24 km, compared to £14 if spent in a supermarket.

We recently experienced the consequences of following a neo-liberal policy in opening up agricultural marketing to the corporate sector. Corporate houses grabbed wheat and other produce from farmers by paying them a little more than the minimum support price (which is deliberately kept low by the government), stockpiled it and jacked up market prices, forcing the consumer to pay through the nose.

The unilateral opening up of the retail sector may also lead to other severe consequences. The farm sector is likely to be affected by cheap imports of highly-subsidised agro products. Global chains, once they set up operations in the country, will find it easier to source cheap products from the world market and sell it in this country.

This will make the local farmers' produce uncompetitive. Most developing countries including India have phased out quantitative restrictions (QRs) on imports. The only way to protect farmers now from cheap, subsidised imports is the tariff mechanism. Off and on demands are being made by the developed world that the developing countries should also effect a corresponding reduction in already low tariffs. The issue of evolving a mechanism to protect some of our sensitive products is still under discussion.

Another possible consequence is the dumping of cheap substandard products. There is no effective quarantine and sanitary and phytosanitary mechanism in place in this country. India doesn't even have a mechanism to test traces of genetically modified organisms (GMOs) in food. India hasn't yet approved any GM food crop. But now even unapproved GM foods are entering the country without adequate checks.

The advocates for opening up the retail sector argue that supermarket majors will help increase food quality norms. This may not be the case. For example, take our recent experiences with the soft drink majors, Pepsi, Coca-Cola and producers of bottled drinking water.

The soft drinks and bottled drinking water were found to be contaminated by pesticide residues, much higher than the permissible limits in EU and other developed countries.

The soft drink majors and bottled drinking water producers later said it would be difficult to implement EU norms in India. It was only then that national norms for bottled drinking water and soft drinks were revised as per global standards and made mandatory.

Source : Outlook Magazine; Dec 31, 2007;[http://www.outlookindia.com /article.aspx?236374]

33

Scrap Special Economic Zones & Promote Agri. Export Zones

Governments oft-repeated mantra for ensuring food security and well being of the farmers has turned out to be a lip service-only to gain political mileage. It's real intention is clear - to benefit big corporate houses and multinational corporations at the expense of farmers. With this intention the government has begun the process of acquiring prime farmlands from farmers at a platter and gifting it to the corporate houses to set up their kingdoms in the name of Special Economic Zones (SEZs). This process can be rightly termed - "Robbing Peter and Paying Paul".

Farmer has become an insignificant being in the eyes of the government - he deserves to be looted and driven to an extreme point of committing suicide. But who needs to be pampered more than a son-in-law? Not the farmer who provides food security, but big corporate houses and the MNCs.

Mighty Kingdoms

According to the recent data put up on the official website 234 SEZs have been formally approved and 162 SEZs have got in-principle approvals. The figures show that 396 mighty kingdoms are coming up in the form of SEZs. There are many more in the pipeline. SEZs are no less than kingdoms. The government has done its best to give SEZs the status of kingdoms, except they will issue their currency notes.

These SEZ will be duty free zones - complete exemption from excise duty, custom duty, sales tax, octroi, mandi tax, turnover tax, as well as income tax holiday for ten years are some of the inducements. They can invite 100 per cent foreign direct investment, enjoy exemption on income tax on infrastructure capital fund and individual investment, and have assurance for round-the-clock electricity and water supply. The SEZ promoters have also been given a waiver from carrying out an Environment Impact Assessment.

Dr.Krishan Bir Chaudhary addressing farmers meeting against Land Acquisition, Saidpur Chowk Distt- Sonipat (Haryana) - 6 June 2011

No Inspection

SEZ owners are authorized not to permit any inspection of their premises by any official of the government and for conducting search or seizure operations without prior permission. They are empowered have their own private security system. By all these qualifications SEZs are, therefore, kingdoms within the Republic of India. What more the government has doled out to the SEZs ? They are permitted to external commercial borrowings up to US $ 500 million without any maturity restrictions and hedge in commodity exchanges. They are free to bring in export proceeds without any time limit and make foreign investments from it and are exempted from interest rate on import finance.

They are allowed to set up off-shore banking units with income tax exemption for three years and subsequently 50 per cent tax for another two years are some of the financial enticements. And if they were to sub-contract production to local manufacturers, there would be duty drawbacks, exemption from state levies and income tax benefits. The fiscal sops extended to SEZs would cause a revenue loss to the government to the tune of Rs 10,00,000 million as per most conservative estimates.

SEZ Forever

Moreover the SEZs are left to use the land gifted to them they way they like. As per SEZ rules, they can use only 35 per cent of the land for the business for which it was allotted, the remaining 65 per cent of the land can be used the way the SEZ owner like - for real estate or for their leisure and pleasure. If the SEZ owner fails in his business operation, the land cannot be restored to farmers.

It would be gifted to another corporate house or MNC who can promise better business. Once a SEZ is always a SEZ, says the law. The official website says that all the SEZs are not set up on farmlands, this may be partially true. But there are instances where SEZs have grabbed large chuncks of prime farmlands and lands owned by tribals.

In Orissa the state government is planning to amend the Scheduled Area Tribal Immovable Property Act to make it possible for big companies to acquire lands owned by tribals. West Bengal has seven approved SEZs and fourteen SEZs which have in-principle approvals. The Kerala has ten approved SEZs and two SEZS approved in-principle.

SEZs kingdoms have come up or slated to come up in several parts of the country including Andhra Pradesh, Chhattisgarh, Delhi, Goa, Gujarat, Haryana, Himachal Pradesh, Jharkhand, Karnataka, Kerala, Madhya Pradesh, Maharashtra, Orissa, Punjab, Rajasthan, Tamil Nadu, Uttar Pradesh, Uttarakhand, West Bengal and in Union Territories like Chandigarh, Dadra & Nagar Haveli, and Pondicherry.

The official website says that for setting up of 396 SEZs 1750 sq km land would be required - all would not be on farmlands. Farmland would constitute only 1393.53368 sq km. Our estimate, however, points out that the loss of farmlands would be much more. The government claims that with an estimated investment of Rs 53561 crore (Rs 535610 million) by 100 notified SEZs, 15,75452 additional jobs would be created - a complete hoax.

The SEZs have not been able to create the desired level of employment generation with the level of investment they have already made. Has the government ever estimate the loss incurred to the nation on account of SEZs ? The food security would be a problem with the shrinkage of farmlands. The displaced farmers would lose their livelihood.

As per the National Sample Survey Organisation (NSSO 2005), the average income of a farming household stands at Rs. 2,115 per month (income from cultivation - Rs. 969; farming of animals - Rs. 91; wages - Rs. 819; and non-farm business -Rs 236). Of these, income from the first two sources (Rs. 1,060) will be immediately lost.

Therefore, each farming household will lose Rs. 12,720 every year. The total loss of annual income for the 1.14 lakh (11.4 million) displaced farm families works out to Rs.145 crores (Rs 1450 million).

Economic Security

As per the National Rural Labour Commission, an average agricultural worker gets 159 days of work in a year; and as per NSSO (2005), the average daily wage of agricultural labour in rural areas is around Rs. 51. Considering this, the estimated 82,000 agricultural labourers' households will lose Rs. 67-crore in wages.

And put together, the total loss of income to the farming and the farm worker families is to the tune of Rs. 212-crore (Rs 2120 million) a year. For the marginalized, the loss of income - even if it hovers around the poverty line - has disastrous implications. Farmland is the economic security for farmers and farm labourers.

What is the way out?

According to the latest data for 2006-07, the new generation of SEZs could generate only Rs 9301 crore worth of exports. Comparatively, the 60 Agri Export Zones which do not enjoy any special fiscal sops, despite putting up a good performance, failed to top decision makers' priority list. The scheme for setting up AEZs was conceived in 2001 and today they are 60 in numers, spread across 20 states.

Despite low investments and inadequate infrastructure, AEZs have received an exports earning of over Rs 60,000 million in the last five years. And in the last six years, not much investment has flowed in. This is despite promises made to agriculturists and traders. Both central and state governments have not been playing a proactive role to bring in investment, let alone encouraging private sector to invest.

As per initial criteria, investments by the Centre, states and the private sector has to be in the ratio of 1:1:2. Accordingly, the total investment for 60 approved agri export zones (AEZs) was estimated at Rs 17,179.50 million. Against this, the total flow of investment to date is only 8,111.80 million. Despite low investments, AEZs could achieve about 50% of the export target (Rs 118, 214.70 million) over a period of five years. The government also admitted that there is a lot of under-reporting by the state governments about the movement of produces from AEZs for exports. The total export figure would be much more than Rs 51, 852.30 million in five years. This should exceed Rs 60,000 million.

Deliberate Negligence

Against the deliberate negligence of AEZs, the government pampered the controversial special economic zone (SEZ) scheme by extending all possible sops. Another reason for exports performance being below target is that all AEZs were not set up in 2001. Many of them were set up much later. And majority of the exporters are of the similar view.

They believe that majority of the investments done so far are by the private sector. The investment could have been much higher had the central and state governments developed better infrastructure, encouraged investment and put in their share of the investment.

According to commerce ministry sources, the APEDA had recently asked for Rs 2,500 million to support AEZs under the government's scheme for assistance to states for infrastructural development for exports (ASIDE). But the ministry agreed to render only Rs 500 million to AEZs under ASIDE scheme. Unlike the SEZs, the AEZs do not have specified physical boundary.

They are confined to specific regions in states, known for growing specific crops. The AEZs are designed for bringing integrated development of larger area including boosting income prospects. So far, the AEZs have been set up for crops like Pineapples, Litchi, Potatoes, Mangoes, Vegetables, Darjeeling Tea, Gherkins, Rose Onions, Flowers, Vanilla, Basmati Rice, Medicinal and Aromatic Pants, Grapes and Grapevines, Kesar Mangoes, Onions, Pomegranate, Banana, Oranges, Mango Pulp, Chilli, Apples, Walnut, Garlic, Seed Spices, Wheat, Lentils and Gram, Cut Flowers, Cashewnuts, Honey, Sesame Seeds, Cherry Pepper, Ginger, Coriander and Cumin.

Revenue Loss

Unlike SEZs, the AEZs do not enjoy any special fiscal sops and hence there is no revenue loss for the government. The government has already admitted that the revenue loss due to SEZs would be over Rs 10,00,000 million by 2009-10. SEZs are being set up on prime farmlands at the expense of food security. Out of the acquired land for SEZs, only 35% is for real business and the rest is for real estate. AEZs are much better for farmers.

AEZs do not displace farmers, rather are aimed at strengthening their income and livelihood. If the government is interested in integrated rural development, it should support AEZs and scrap the SEZ scheme.

If SEZs are to set up it should be done on 552, 692.26 sq km of identified wastelands in the country.

Source : Economic Journalist Magazine; July, 2007 & Global Information Media; October 3, 2007; [http://globalcommunitywebnet.com/GIMProceedings/gimLetter KBChaudhary.htm]

34

Government cannot abdicate its Responsibilities

Prime Minister Manmohan Singh and the National Development Council (NDC) have failed to realise the problems facing Indian agriculture and the farming community. Mere announcement of a food security mission to increase the production of wheat, rice and pulses, and pegging up the Central government's investment to Rs 25,000 crore within a span of four years will not solve the problems. The figures and words in the NDC resolution sound great.

But what does it means to farmers? The PM, in his opening speech, raised concerns over farmer suicides, but he did not dare to say that he had resolved the issue in Vidarbha through a package he had announced.   Incidences of farmer suicides in different parts of the country are on the increase. Increasing public investment in agriculture is a right approach, but at the same time it should be ensured that the money is rightly used. Imperfections in government machinery should be immediately removed to ensure the success of any development and welfare scheme.

The PM has, on several occasions, admitted to corruption and shortcomings in government machinery. But why is he not prepared to set a definite time frame for the eradication of corruption? Instead, a case is being made out that the government machinery cannot be efficient, welfare and development schemes should be scrapped as they are wasteful public expenditure and there should be greater involvement of corporate houses in agriculture for the benefit of farmers. Corporate objectives are to garner greater profit, while the priority of a truly welfare state is development. Unfortunately, the government of the day is abdicating its responsibility.

The recommendations of one of the NDC sub-committees suggest facilitating greater corporate involvement by way of contract farming and corporate farming (of course, with a rider: only when the country attains a reasonable level of food security). The so-called wise men in the country should know that the distress of cocoa growers in Africa and profitability of multinational companies are on account of contract farming. This is just one of several such instances across the world.

What according to the NDC sub-committee is the "reasonable level of food security at both macro and household levels" in the country to allow corporate farming? This is not clear, but indications are there that corporate farming would soon become a reality. The process has already begun by inducing farmers to lease out their land to joint ventures-land share companies-to be formed in association with existing agri-processing companies.

Instead of giving more directly to farmers, the government is intending to take away whatever they get. Fertiliser subsidy is routed to farmers through fertiliser companies. According to expert studies, less than 40% of the subsidy reaches the farmers. The concerned NDC sub-committee suggested restructuring the fertiliser subsidy and diverting a substantial amount of it to research, marketing and infrastructure development.

It says that diversion of fertiliser subsidy to research, marketing and infrastructure would be "WTO compatible", as if our existing level of fertiliser subsidy is not WTO-compatible. The real intention is to fund MNCs undertaking research in the transgenic agriculture. Why can't the government think of routing fertiliser or any other intended subsidies directly to farmers? In Europe and North America, farmers get direct subsidy. Does the PM or the NDC mean that Indian farmers cannot be trusted should not be given direct subsidy?

The PM has criticised small holdings as a drag on farm productivity. Does he not know that countries like Japan and China have farm holding sizes much smaller than that in India, and yet they have high farm productivity.

Food security and agricultural productivity can only be ensured if farmers are taken into confidence. Farmers should be left alone to plan farming and the intended benefits should go directly to them.

Source : The Financial Express; Jan 11, 2007;[http://www.financialexpress.com/news /Govt%20cannot%20abdicate%20its%20responsibilities%20/136690]

35

Market access in Farm Negotiations

Market access is one of the three important pillars for farm negotiations. Access to markets in the developed world for products and produces of developing countries is vital for farmers of the Third World countries getting justice in global trade. Farmers in the third world countries have been worst hit from global trade caused by heavy subsidies and support given by the developed countries to their farmers.

The developed countries have not only blocked market access to the developing countries but also are bent on pushing the heavily subsidised farm products into the market of the developing countries. Thus the farming systems in the Third World countries are extremely vulnerable.

The heavy domestic subsidies and support coupled with export subsidies and credits rendered by the developed countries have depressed global prices resulting in Third World farmers not getting remunerative prices. Therefore, the market access cannot be judged only on the basis of tariff and non-tariff barriers.

The other two pillars of farm negotiations have an effect on market access. Elimination of all forms of subsidies is a must for rendering market access. As per several studies it is a small group of rich farmers in the industrialised countries who are the real beneficiaries of the subsidy regime.

The poor farmers with smaller landholdings in the industrialized countries are denied of the benefits being enjoyed by the rich farmers having large land holdings. I extend my sympathy to all the poor farmers of the world. Majority of the poor farmers live in the developing countries and very few of them are in industrialised countries.

According to FAO estimate, more than 2.5 billion people in the developing world are dependent on agriculture for their livelihoods. Agriculture is not only a means of sustenance for them, it is a way of life. Most of the 842 million under nourished people in the developing world today are from farming families.

Over the past decades the models adopted by the US and the European Union have pushed majority of the farmers out of agriculture. Small number of big farmers are now in agriculture in these countries.

This has resulted in large mechanised farms controlled by few rich farmers and agri-business corporations. Agri-business Corporations now command agriculture directly or indirectly in these countries.

I caution the policymakers, that if such models are adopted in India and other developing countries it would spell disaster to millions of farming communities and cause massive migration to cities increasing the numbers of slum dwellers. The unemployment problem will assume unresolvable dimensions.

In this country there are strong advocates of the western model of large mechanized farming and corporate farming. I ask such proponents of truncated farm practices "Why there are 17 million poor in US living below the poverty line? Has not the number of poor increased in US due to forced migration to cities?"

We have examples of exploitative corporate farming in India in large tea estates and plantations. Why are the profitability of tea companies eroded? Why are the indices of these tea companies dipping in stock markets? Why tea exports, today, are on a declining trend? Why is the maintenance of estates poor, particularly relating to soil health?

These things are happening despite receiving much patronage

from the government. The same is true for other large plantation estates. Is it not a failure of corporate farming?

The Third World countries are being time and again insisted upon to open up their market to the heavily subsidised products and produces of the developed countries.

The World Bank and the IMF have become instruments to pressurize developing countries to open up their markets. The pathetic state of farmers in sub-Saharan Africa is the result of such opening of the market.

Unilateral trade liberalization undertaken by sub-Saharan Africa at the instances of the World Bank and IMF has cost these poor nations USD 272 billion over the past 20 years. And these losses dwarf the USD 40 billion worth of debt relief agreed to by G-7 finance ministers.

Two decades of forced liberalization has cost sub-Saharan Africa roughly what it has received in aid. Effectively, this aid did no more than compensate African countries for the losses they sustained.

The doubling of aid to Africa by 2010 by about USD 25 billion a year as promised by G-8 will further push this continent into a debt trap ! Aid is no solution; forced unilateral trade liberalisation of sub-Saharan Africa coupled with distorted global trade is the cause of distress.

Parallel to this, India removed all quantitative restrictions (QRs) on imports. We now have tariffs as mechanisms to deter influx of cheap imports. We are now being told to reduce our tariff levels.

Our bound tariff rate on soybean oil is only 45%. This is too low. Already, vegetable oil imports are rising and oilseeds cultivation is not picking up to the desired extent. The Technology Mission could achieve almost self-sufficiency in oilseeds production by 1996-97.

Now we are importing over 45 percent of vegetable oil for meeting our domestic needs. The foreign exchange outgo on this account is over US$ 1800 million. Cultivation of pulses is not increasing to the desired extent and the imports of pulses are also rising.

The cascading effect of depressed global prices resulting in low returns to farmers is being felt in India. The suicides among farmers have become rampant. According to the latest report of the National Sample Survey Organisation (May 3, 2005) nearly 48.6 percent of the 90 million farm households are caught in debt trap.

In Andhra Pradesh, 57 out of 100 indebted households are beholden to moneylenders. The African cotton growers are already suffering on account of depressed global prices.

The cocoa growers in Africa are suffering due to low prices paid to them by multinational corporations, which have gained excess to market this produce. These multinationals procure cocoa from African growers at cheap rates and sell coffee in global markets reaping huge profits. This is a glaring example how contract farming can work against the growers' interests.

Statistical data compiled by FAO show developing countries have turned from net exporters to large importers of food from developed countries: a food trade surplus of US$ 1 billion in the 1970s was transformed into a deficit of US$ 11 billion in 2001.

Subsidised exports from developed countries make imported food cheaper than local products in developing countries which undermines local agriculture.

Also, the overall decline in support to agriculture in developing countries has worsened the situation. The disastrous deterioration of the terms of trade (rationalisation of prices of exported goods to the price of imports) for the LDCs has resulted in "a consequent transfer of income from developing to developed countries."

The farmers in the developing world are suffering from lack of market access. The developed world continues to maintain high tariff barriers and often resorts to tariff escalations.

According to FAO the average tariff for agricultural products in developed countries is 60 percent, compared with an average 5 percent for industrial goods. These tariffs on imports by developed countries are unfair to developing countries, which are highly dependent on the export of agricultural commodities.

Furthermore, exports from developing to developed countries face tariff escalation: higher tariffs are levied on goods exported at more advanced stages of processing. Tariff on fully processed foods in many cases are more than double the tariffs on basic food commodities.

Sanitary and phytosanitary (SPS) measures and technical barriers to trade (TBT) are often used politically as weapons to stop imports from the Third World countries.

I am not against SPS measures as such if it is used genuinely to protect the health of the citizens and are based on scientific evidences. It should not be used as a political weapon to deter trade. In May this year U.S. rejected 251 food and drug consignment of 54 leading export houses of India citing various quality and technical parameters.

I am constrained to say that the Indian Government has not taken up this issue seriously to find out as to whether the rejection has been made on scientific grounds or to deny exports from India.

The European Union continues with its programmes of rejection of Indian consignment on SPS grounds. SPS and TBT are important in relation to market access.

With great difficulty the developed countries agreed in early May in Geneva on the process of converting specific farm tariffs based on quantities imported into price-based ad valorem equivalents (AVEs) with the exception of sugar.

Many developed countries have high specific tariffs based on quantities of farm products imported in lieu of ad valorem duties, while developing countries in general have ad valorem tariffs.

India has ad valorem tariffs on all farm produces, with the exception of almonds. Now after the agreement to convert specific farm tariffs based on quantities imported into AVEs, there is a need to move to Stage-II i.e. specifying how tariff reductions would be made.

Unfortunately, the July 2004 package, though mentioned a three-tiered formula for tariff reduction, it did not clearly mention how reductions would be made under different tiers.

There are already two formulae for tariff reduction - one is the Swiss formula and the other is the Uruguay Round formula. In the last month in Geneva, the farm talks could not resolve this issue.

Only at the Dalian mini-ministerial the US and EU have reportedly accepted a compromise formula drafted by G-20 on tariff reduction. The G-33 has supported this proposal. However, Switzerland and Japan have opposed the G-20 proposal.

The G-20 has proposed 5 bands for developed countries and 4 bands for developing countries for tariff cuts. The greater number of bands for developed countries aims at implementing progressively and taking into account the wider dispersions of tariffs.

In this context I would say that the Swiss formula where high tariffs would undergo higher cuts than lower tariffs, should be applied to developed countries while the Uruguay Round formula which gives more flexibility in tariff reduction should be applied to developing countries. The least developed countries (LDCs) and the new members of WTO should not be forced to further tariff reductions.

This would result in assuring a level playing field to a certain extent. The tariff reduction formula should be negotiated before addressing the issue of flexibility.

The developing countries should effect tariff reduction only after the developed countries substantially reduce their domestic and export subsidies and support. I would say all subsidies and supports are trade-distorting and hence should be totally removed.

There should be a cap to bring down high tariffs. In this context the caps should be different for developed and developing countries. G-20 has fixed 100% for developed countries and 150% for developing countries.

Mechanical approach for determining thresholds should be avoided as developed countries have highly skewed tariff structures. There should be an additional formula for reducing tariff escalation in developed countries and the products should be identified.

The tariff reduction in farm commodities are hoped to be

achieved. But one concern still remains for a specific commodity of India's export interest. This is sugar. The developed countries, particularly the EU have not agreed to convert their specific duty on the quantity of sugar imported into AVEs.

In EU the duty on sugar is over 250% and India being a sugar exporter finds it difficult to enter European markets. The only way is through tariff rate quota (TRQ) which allows an export of only 10,000 tonne a year against concessional duty.

This quantity under TRQ is too low. There is a need to ask EU to convert its specific duty on sugar into AVEs and effect a drastic reduction in its sugar tariff. Expansion of TRQs by the developed countries should not be viewed as an act of charity.

Already these countries have high tariff barriers. The most important thing is the need for drastic reduction of high tariff barriers. The developed countries should provide duty free and quota free access to the produces from LDCs if they are really eager to help them, rather than extending aid.

The developed countries should be allowed to designate only a limited number of sensitive products and it should be credible and reasonable in relation to trade. If the designation of sensitive products causes deviation from tariff reduction formula, it should be compensated by expansion of TRQs.

On the other hand the developing countries and the LDCs should be allowed to designate their special products without any prescribed limits. Designation of Special Products and availing of Special Safeguard Mechanism should be an integral part of Special and Differential Treatment for developing countries and the LDCs.

The developed countries have not fulfilled their commitments to reducing their subsidies and all supports in the first phase. In place of reducing, they have increased their subsidies.

In a situation like this, how the developing countries could compete in the market? First and foremost the developed countries need to fulfill their own commitments in accordance with AOA.

In the present circumstances, India has become a dumping ground of heavily subsidized agricultural products. Distorted price of all agriculture products have been damaging and squeezing the market for the developing countries.

Unless the developed countries totally scrap their subsidies and all supports, the question of tariff bound reduction by India does not arise. India must reimpose quantitative restrictions (QRs) to protect the food security of the country and safeguard livelihood concerns of the people.

Source: The paper presented in "Pre Hong Kong Ministerial Consultation: Agriculture Negotiations", jointly organized by Ministry of Commerce & Industry (Govt. Of India) and UNCTAD at New Delhi on 19th and 20th july 2005.

International Issues

36

Global Civil Society Opposition to the "Pledge Against Protectionism" in WTO

$\mathbf{A}$ wide variety of civil society experts from the global present in Geneva for the 8th Ministerial meeting of the World Trade Organization (WTO), voiced their opposition to the idea of a standstill on tariffs in the WTO proposed within the "Pledge Against Protectionism" . The present policies are unfair and unbalanced and favor trade concentration in the hands of few trans-nationals corporations for exports and are based on the growing and irrational exploitation of human and natural resources from the developing countries, and the developing countries are facing negative impact of trade that deepen poverty & inequality.

We condemn the double standards of countries that proposed an additional pledge against protectionism. While, the developed countries resorted to heavily protecting their agri - business corporations, it is travesty of justice to call upon the developing countries and LDCs to remove the minimum support structures created to safeguard the livelihoods of millions of small and marginal farmers.

Without the removal of Green Box subsidies (a great protection given to agri - business corporations by the developed countries), the developed country demands in this pledge that developing countries and least developed countries should not be able to exercise their rights under the WTO to raise their applied tariffs to their bound rate is unfair. Agricultural liberalization in India has already caused 256000 Indian farmers to commit suicide. We call on India not to sign this pledge or agree to any standstill at the WTO.

There is nothing new from what these developed countries are saying about protectionism. They have to deal with their domestic protectionist tendencies first before they ask for multilateral cooperation. These are double standards at their worst. The developed countries' pledge to fight protectionism has no shame at all. They have no moral ground to talk against protectionism when they are the worst culprits.

This is a desperate attempt to cover up the truth that the rotten and bankrupt "development" paradigm of so-called "free market" globalization is the main cause of the global economic and financial crisis.

While developing countries have been facing the challenges arising from a global economic crisis they did not cause, the "pledge" that is being promoted in the name of fighting against protectionism will bring in restrictions on using multiple policy tools that these countries have fought throughout the WTO negotiations to save, which are a right that WTO law should protect and not jeopardize, and which are essential to any development prospects in these countries

The global trade framework must provide member countries sufficient policy space to pursue a positive agenda for development and job-creation. Trade rules must facilitate and not hinder global efforts to ensure genuine food security, sustainable development, access to affordable healthcare and medicines, and global financial stability.

Source : Editorial; Kisan Ki Awaaz; Jan, 2012; [http://www.kisankiawaaz.org /Global%20Civil%20Society%20Opposition%20to%20the%20%E2%80%9CPledge%2 0Against%20Protectionism%E2%80%9D%20in%20WTO.html]

37

Global Contamination

The world is facing contamination of natural environment from two sources. One, the uncontrollable release of radioactive contaminants from the crippled Fukushima Nuclear Power Plant and the other from illegal planting of Genetically Engineered seeds worldwide by criminal corporations like Monsanto, Bayer, DuPont, Syngenta and Dow Chemicals.

The four reactors at Fukushima Nuclear Power facility, owned by Tokyo Electric Power Company [TEPCO], were knocked out of operation by the 9.1 earthquake off the Japanese coast that caused tsunami waves. The waves killed over 18,000; thousands remain unaccounted for. The damage was so severe that it has virtually knocked out the supply chain of Japanese industries including major car manufacturers.

The tsunami waves damaged the cooling system of the reactors that started uncontrollable meltdown process. That led to release of lethal radiation in the form of Strontium-90 and Iodine-131 that has been found at high enough levels as far as France and now covers the entire United States. The radiation spread all over the world by 15th March and India is also receiving high dose. Senior scientists have warned that the situation is so bad in Japan that the entire Tokyo region with 40 million people may have to be evacuated; others are advocating evacuation of the entire Japanese population to safer regions.

The Government of India has already instructed Port Authorities to test all ships originating from Japanese ports because the ships and their consignments will be radioactive. The Japanese accident had to happen. The technology for generating electricity using uranium has time and time again proven to be unsafe. Every nuclear reactor is now a ticking nuclear bomb; similar explosion can happen is any of the 400-odd operational reactors.

It can happen to any reactor located on fault zones and especially those located on the coasts because of vulnerability to natural disasters. Underplaying such possibilities is essentially a public lie. The second source of contamination is coming from illegal planting of Genetically Modified seeds. In an asinin example of bravado, one of GM seeds companies, had the temerity to plant GM corn adjacent to Bihar Chief Minister's ancestral farm in Bihar.

When he was told that GM maize has been planted he spoke with Sharad Pawar to stop this sort of illegal planting. Since Sharad Pawar had been insisting on allowing GM crops to be planted, Nitish Kumar called up Jairam Ramesh and asked under which Ministry comes Genetic Engineering Approval Committee.

Despite Jairam Ramesh directive to stop this sort of illegal planting, how GEAC is conducting the trials right under the nose of powerful Chief Minister remain shrouded in mystery. The GEAC and the entire group of scientists working for the GM seeds multinational corporations are pushing for a new GM technology bill against the wishes of majority of farmers and the wishes of the Indian society.

Contamination from GM seeds is irreversible; no technology exists to decontaminate natural seeds. This sort of large scale contamination is a proven fact and it is a deliberate act on part of seeds corporations to contaminate every natural seeds worldwide.

They don't want to leave any option to consumers who would like to eat natural foods grown from natural seeds. Such irresponsible behaviour amounts to subjecting world's population to the largest biological experiment without our consent.

And this biological experiment, that will destroy the natural environment and human and animal health, the Government of India is actively colluding.

The two sources of contaminations one radiological and the other biological- have the property of altering natural genetic structure of all living things. Both are known to cause degenerative diseases and environmental destruction on massive scale.

The two events show how science itself has been enslaved to serve corporate interest.

Source : Editorial; Kisan Ki Awaaz; April, 2011;[http://www.kisankiawaaz.org /Global%20contamination.html]

38

International Meet Ask for Right to Food

Slow Food international organized the meet TERRA MADRE - 2010 in Turin (Italy) from 21 to 24th October 2010 and more than 5000 delegates participate from 154 countries. They believes that the development of sustainable food policies and the enactment of legislation in this sphere must be firmly rooted in a human rights based approach, which envisages the right to food as a predominant component.

Dr. Krishan Bir Chaudhary & Mr. Carlo Petrini, President, Slow Food International in 21-24 Oct. 2010, Turino, Italy

The right to food can be defined as the right of every person to have regular, permanent and unrestricted access, either directly or by means of financial purchases, to quantitatively and qualitatively adequate and sufficient food corresponding to the cultural traditions of the people to which the consumer belongs, and which ensures a physical and mental, individual and collective, fulfilling and dignified life free of fear.

They embraced the foregoing definition and recognized the corresponding obligations of states, which are called upon to respect, protect and fulfill the right to food and highlight the vital role of the concepts of food security and food sovereignty as guiding principles for all state action at the community, local, national, regional, and international level. They firmly believe in the customary nature of the right to food under international law and urges states to ratify all relevant international and regional instruments and calls upon states to explicitly include the right to food in the national Constitution or analogous instruments thus imposing on all branches of the state on obligation to take measures to respect protect and fulfil the right to food. They identified and made recommendations with regard to a number of areas requiring both legislative and policy interventions and urges States to improve access to the sustainable use and allocation of resources such as land, water, labour and genetic resources among all segments of the population with no discrimination. They urged states to shift the structuring measurement allocation and disbursement of subsidies from the current commodity-heavy and industrial agriculture focus to local small scale sustainable and/or organic community or family based farming and food harvesting.

Considering the negative effects trade liberalization may have on the enjoyment of the right to food, Terra Madre encourages states to work towards ending their dependence on international trade and to increase capacity to produce the food they need locally and nationally and urges states to regulate and monitor transnational corporations in order to limit their negative impact on the enjoyment of the right to food. They urged international financial institutions to act in compliance with human rights principles and refrain from adopting any policy or program that violates or has negative repercussions on the right to food. Terra Madre is persuaded that plant genetic resources constitute a common heritage they should not be subject to commercial patenting and should be shared among farmers worldwide. They acknowledged the vital role of farmers and indigenous communities and calls on states to recognize their rights. States must further ensure that the development of intellectual property rights regimes is compatible with and instrumental to the realization of the right to food.

Source : Editorial; Kisan Ki Awaaz; Nov, 2010;[http://www.kisankiawaaz.org/International%20Meet%20ask%20for%20Right%20to%20Food.html]

39

Opening up Indian Food Trade
to EU & US

The decision of Government of India to give Free Trade status to European Union in agriculture products spells disaster for Indian farmers. If the ostensible purpose of the Indian Government is to ensure food security and food sovereignty, then this move will destroy our farmers and our food growing potential.

The main causes for food insecurity and consequential widespread malnutrition in India is not inability of farmers to grow enough food. Farmers have been led to believe that: (i) they must earn cash to raise their standard of living; (ii) in order to earn cash they must produce for the market, (iii) earn cash income and (iv) use the cash to buy food and other services from the enforced market economy.

In the US and EU a farmer is fortunate to get a net return of 5% on revenue; in the LDCs the entire household must work, including women and children, to earn a subsistence or starvation wage. Indian farmers have been duped by seed, fertilizer and pesticide companies ever since the Green Revolution started.

Market forces entice farmers to grow cash crops that are raw materials for food, feed and fibre industries controlled by global cartels. They are lured by promises of higher income but it falls when there is bumper crop and it falls when yields drop. This is of course Government engineered. The major threat to Indian agriculture would come from the huge subsidy that the European Commission and the US government provide to its biggest farmers [not their own small farmers, please note] that allows the cartel backed by their respective governments to manipulate global agricultural commodity prices exactly as they want.

The top 10% of the biggest agricultural producers [in the USA] received more than 72% of its $23 billion subsidy programs in 2005. Meanwhile, 60% of all US farmers do not collect any government subsidies.

Dr. Krishan Bir Chaudhary, President, Bharatiya Krishak Samaj, addressing UNI Global Union (Switzerland) meeting on Walmart Global Track Record and the Implication for FDI in Multi-Brand Retail on 20 March 2012 in Constitution Club, New Delhi

The European Commission data show that in 2004, US$36 billion (€28.2bn) of direct subsidies was paid out of a total Common Agricultural Policy (CAP) budget of $58bn (€45.6bn) and 7% of Europe's primary food producers received more than 50% of these payments.

The biggest 2,460 farmers in Europe received on average $667,000 (€524,000) each, totalling $1.7bn (€1.3bn). The sops totalled over 140 billion dollars over 1995-06 in the US as well and covered cotton, canola, soy, sorghum, among other agriculture commodities.

These products would be dumped on India by the combined might of US and EU, while their corporatized farmers enjoy huge subsidies. The US Government is on record for stating that food shall be used as a weapon. Take rice, which is staple food for nearly 3.7 billion Asians.

The US Government provided about one billion dollar subsidy to just three rice growers in the US over 1995 to 2006: $526 million to Riceland Foods, $314 million to Producers Rice Mill and $146 million to Farmers' Rice Cooperative.

And US rice is contaminated with genetically engineered genomes that would kill healthy Indians. When German farmers found that genetically engineered seeds had contaminated their foods, they requested their Government to enhance contamination level from 0.1% to 0.9%, a nine times increase.

In matters of agriculture, the Government of India must stop Free Trade with EU and the US because in these two regions small farmers have disappeared, industrial farming is the norm and that is controlled by five seeds corporations and about ten agri-business corporations.

Together they control global supply of food grains, milk, meat and all foods. The Government of India must not support global monopolies if it has any respect for law, morality and socio-economic reality of India where 70% household are dependent on farming.

Source : Editorial; Kisan Ki Awaaz; July, 2010;[http://www.kisankiawaaz.org /Opening%20up%20Indian%20food%20trade%20to%20European%20Union%20and

40

EU - India FTA will Damage Indian Farmers

The present global trade regime does not present a fair play. Developed countries are backing their farm production with a heavy dose of subsidy and are also protecting their farm sector by high tariff barriers and unjustified non-tariff barriers. In this, situation if India reduces its bound tariff rates (which are already lower), there would be an influx of cheap subsidized imports from the developed countries. India, being a tropical country, is bearing the brunt of climate change. We are facing extreme weather conditions leading to droughts, floods, and cyclones.

Indian agriculture has become vulnerable not only due to climate change but also due to the policy of economic liberalization and globalization of trade pursued by the government. Agriculture in India is a way of life of 750 million small and marginal farmers. It is a source of livelihood and not a business. In fact, agriculture cannot be a business for family farms; it has to be a source of livelihood. This truth holds good also in developed countries of Europe and the Americas, where small family farms still exists.

History has shown that small family farms did not depend on subsidies. The subsidy regime began with corporate and mechanised agriculture which has to depend upon a heavy dose of subsidy. Thus agriculture trade liberalization without removal of all subsidies (Amber, Green and Blue Boxes) will lead to cheap subsidized imports in the developing and least developed countries. The Indian farmers will be largely affected by such cheap imports. Their livelihood would be at risk. Therefore, there can only be two alternatives One is removal of all subsidies and a meaningful tariff regime and removal of unjustified non-tariff barriers, including sanitary and phytosanitary norms.

Indian farmers can survive without any subsidy, provided the rules of the trade are fair and equitable. The other option is to have a protected market regime having the options to impose quantitative restrictions on imports and exports. As long as unfair trade practices remain and corporate houses control agriculture and agriculture trade, the world will always witness volatility in prices. Volatility in prices would be caused if land is diverted for production of bio-fuel crops and food crops are used for production of fuel oil. Such was the case in 2008 when the food prices soared and moved in tandem with the prices of fossil oil. The FAO then raised the alarm of likely global food insecurity. Liberalising investment restrictions in retail will lead to the loss of livelihood of millions of people dependent on retail business. In India, domestic corporate houses have entered in retail business and farm produce trade. We are experiencing the results of high retail prices.

There is a wide divergence in wholesale and retail prices. This evidently points to the deliberate manipulation in trade. The future trading has also worsened the situation. The government has banned futures trading in select agriculture commodities to rein in the rising prices. This present situation of rising prices does not benefit the farmers as the prices fall at the time of harvest. The price rises when the produces are in the hands of the trade and industry. The consumers suffer on account of rising prices. Opening up for investment in the processing sector will strengthened the corporate stronghold over agriculture. The ideal solution would be to involve unemployed rural youth in processing through proper incentives and support.

Source : Editorial; Kisan Ki Awaaz; March, 2010;[http://www.kisankiawaaz.org /EU%20-%20INDIA%20FTA%20Will%20damage%20Indian%20Farmers.html]

41

Russia: a Friend in need is a Friend indeed

Russia has once again reaffirmed its time tested friendship with India. Understanding the need for a multi-polar world and for collectively tackling global problems like financial crisis, energy and food security and climate change, it has said that India is "a deserving and strong candidate" for a permanent seat in an expanded UN Security Council. Russia has also supported India's full membership in the Shanghai Cooperation Organisation (SCO) and in the Asia-Pacific Economic Cooperation and also called for lifting the moratorium on expanding the APEC membership. Feeling the need for Russia's involvement in the Asia-Europe Meeting (ASEM) India has supported Russia joining that dialogue forum at the 8th ASM scheduled in Brussels in 2010.

These assurances and support to India came in the form of the joint declaration signed by both the countries on December 8, 2009 on the occasion of the visit of the Prime Minister of India, Dr. Manmohan Singh to Russia. The strategic partnership between the two countries calls for "building a new, democratic and fair multi-polar world order based on collective approaches, supremacy of international law and adherence to the goals and principles enshrined in the UN Charter." Russia also assured India in bilateral energy cooperation, including that in the area of nuclear energy, cooperation in meeting the threat of extremely dangerous, infectious and other contagious diseases, counter-terrorism, timely response to natural and man-made disasters and stability in Asia-Pacific, particularly in Afghanistan.

Both sides agreed to work for global non-proliferation and complete and verifiable elimination of nuclear weapons, ensuring international information security and preventing deployment of weapons in outer space. The nature of cooperation and support and the role assigned to India by Russia far outweighs that given by the US president Obama to India during the recent visit of the Indian Prime Minister Dr. Manmohan Singh to that country.

Apart from the traditionally strong cooperation in space and defence, India can benefit from Russia's rich deposits of hydrocarbons and expertise in infrastructure building, particularly construction and engineering. Russia can benefit from India's expertise in pharmaceutical, information technology and communication sectors.

There is natural complementary between the two countries in rough diamond trade. Russia is the largest producer and India is the centre for cutting and polishing. Both sides have agreed to boost bilateral merchandise trade to $ 20 billion by 2015. The friendship between the two countries dates back to the establishment of diplomatic relations in 1947, after India got independence from UK's dirty colonial regime. The multifaceted India-Russia relationship is not influenced by the engagement of these two countries with the rest of the world.

The early foundation of the India-Russia (then Soviet Union) friendship was laid by the first Prime Minister of India, Pandit Jawaharlal Nehru who opted for keeping equal distance from two global power blocs -the United States and the Soviet Union-and co-founded the Non-Aligned Movement. India-Russia relationship was further strengthened by Prime Minister Indira Gandhi who liberated Bangladesh with Soviet Union's moral support.

Source : Editorial; Kisan Ki Awaaz; Dec, 2009;[http://www.kisankiawaaz.org/Russia %20%E2%80%93%20a%20friend%20in%20need%20and%20a%20friend%20indeed.h tml]

42

Time to wind up WTO

After successive failures of a series of attempts to revive the multilateral trade negotiations, it is time to wind up the WTO, says India's Bharatiya Krishak Samaj (Indian Farmers' Organisation). As not much progress in multilateral trade negotiations was made on basis of the drafts issued in May this year, the WTO again on July 10 issued two revised drafts - one on agriculture by Crawford Falconer and other on NAMA by Don Stephenson. There is practically not much difference in spirit of the drafts issued in July and those released in May.

The revised WTO text on agriculture has completely ignored the food security and livelihood concerns of the farmers in the developing countries. It has proposed a weak defence against the possible influx of cheap and subsidized imports by suggesting a complicated system for implementation of Special Safeguard Mechanism (SSM) by the developing countries. In contrast the Special Safeguard (SSG) implemented by the developed countries is simple and effective.

The revised draft, though has the provision for self-designation of Special Products by developing countries, the provision suggested is not enough to protect food security and livelihood concerns. "All crops grown by farmers in our country are linked to livelihood concerns and therefore we should have the right to designate all crops grown by farmers as Special Products. If the WTO is not willing to give us this right along with an effective SSM then it should allow countries to impose Quantitative Restrictions (QR) to check any possible surge in imports," said the President of Bharatiya Krishak Samaj,

Unfortunately the revised draft has not proposed much in the direction of calling for a drastic cut in heavy subsidies in the developed countries and sharp reduction in their high tariff barriers. It allowed subsidization and cross-subsidisation through all avenues - AMS box, Blue Box, Green Box. While allowing developed countries to protect their agriculture the draft has suggested developing countries to open their doors for imports.

Without resolving the issue of livelihood protection and food security of the developing countries in general, the revised farm draft has extended some concessions to least developing countries (LDCs), small vulnerable economies and 'other developing countries' may possibly divide the unity amongst the developing countries, if not tackled effectively. The revised NAMA draft released by Don Stephenson has sought to take away the flexibilities to industries in the developing countries with the introduction of 'anti-concentration' clause.

While the July 2004 Framework agreed that the flexibility cannot be used by developing members to exclude entire HS Chapters, the latest draft suggested restrictions beyond this mandate. Such unduly restrictive clause disregards the realities and sensitivities of the industries in the developing countries. Flexibilities are more needed for the protection of small and medium-sized industries.

The revised NAMA draft still continues to link tariff reduction coefficients with flexibilities. The flexibilities have to be treated on stand-alone basis and there should be no trade-off between flexibilities and tariff reduction coefficients. The latest NAMA draft has also attempted to create a division in the unity of the developing countries by proposing additional flexibilities to some.

The proposal for negotiations in remanufactured goods finds place in the revised draft indicating convergence on this issue, which is far from reality. In the earlier draft, the issue was under the square brackets reflecting lack of consensus on the subject.

The controversial coefficient ranges for developed and developing countries for cutting tariffs through a "Swiss formula" still remains in the May text as also the percentages of tariff lines that can have flexibilities from the full tariff cuts, according to a "sliding scale".

The WTO has failed to fulfill the objectives of the Doha Development Round. This is primarily due to the adamant attitude of the developed world to continue their protectionist regime in agriculture through high subsidies and tariff and ask for market access in the Third World. In this context, Bharatiya Krishak Samaj firmly believes that it is time to wind up the WTO instead of creating a drama of convening a mini-ministerial on July 21.

It would havebeen better for the Indian Commerce Minister, Kamal Nath to say that India could boycott the mini-ministerial rather saying that he would walk out if the WTO fails to protect the interests of small and marginal farmers and infant industries. There is no doubt that Kamal Nath has made his statement about walkout due to the compulsions of political situation in the country and the upcoming polls, rather than his desire to protect the country's farmers.

Source : www.westender.com.au; July 20, 2008;

43

How should India Respond to WTO Drafts

The two new drafts for Negotiations in Agriculture and Industrial goods (NAMA) issued by the WTO on May 19 this year is a mockery of the current situation. The two drafts do not accommodate the concerns of the developing world. The WTO is practically hopeless and helpless after successive failures of a series of attempts to revive the multilateral trade negotiations.

This is primarily due to the adamant attitude of the developed world to continue their protectionist regime in agriculture through high subsidies and tariff and ask for market access in the Third World. Interestingly, trying to encash upon the current food crisis, the WTO director-general, Pascal Lamy instead of urging the developed world to remove the distortions in trade suggested the developing countries to open up their doors.

The chair of the agriculture negotiating committee, Crawford Falconer, while issuing the draft, claimed that it was a result of discussions held since September last year. Similarly, the chair of the Nama group, Don Stephenson, claimed that his revised text was the product of the bilateral and plurilateral consultations he had over the last few weeks and was also built upon the past years of negotiations. He, however, said, "this revised text is another step in the process and might be subjected to further revision."

At least Stephenson is more frank in his admission than Falconer. The two new drafts show very little areas of convergence despite the WTO director-general Pascal Lamy claiming to the contrary. The NAMA draft has completely ignored the Doha mandate of less than full reciprocity in tariff reduction for developing countries. It has linked the tariff reduction coefficients with the flexibilities for developing countries.

If the Doha mandate is to be followed then the issue of flexibilities for developing countries should be treated separately and not in the way Stephenson has done. Stephenson has also sought to divide the developing countries by extending additional flexibility to some members.

The Nama text, if accepted would doom the future of many small and medium sized units and also cause problems to the fishery sector. The farm draft has ignored the very basic aspect of food and livelihood security of the Third World. Though it has recognized the developing countries' right to self-designate their special products based on food and livelihood security and rural development, it has proposed the minimum limit of 8% of the tariff lines and maximum limit of 20% of the tariff lines.

The crucial issue is the maximum extent of self-designation of special products by developing countries which is in square bracket, implying the need for further negotiations. The draft has recognized the developing countries' right to self-designate special products guided by indicators based on the criteria of food and livelihood security and rural development. G-33 countries suggested the minimum limit of 20% for self-designation of special products.

The problem is more for a multi-crop country like India where farmers depend upon a large number of crops for their livelihood security. Thus, capping of special products by a fixed percentage would not be enough to protect the farmers.

The Falconer draft has presented a complex and complicated process for implementing special safeguard mechanism (SSM) to prevent a surge in imports in the developing countries, while the special safeguard (SSC) remains simple for developed countries. The draft does not provide for conversion of all the complex and specific tariffs in the developed countries into their ad valorem equivalent, which is necessary for transparency and effecting realistic reduction. It reflects an uncertainty related to desired reduction in subsidies in the developed world. It only indicates some range of cuts that too within square brackets, implying the need for further negotiations.

Source : The Economic Times; May 27, 2008;[http://economictimes.indiatimes.com /opinion/how-should-india-respond-to-wto-drafts/articleshow/3074723.cms]

44

Dialogue with Mr. Pascal Lamy (DG, WTO)

Basically as an agriculturist and equally concerned as the leader of Indian Farmers' Organization, I have closely watched with utter disappointment and dismay how the W.T.O. has gone into monopoly control of developed countries. OECD countries are aggressively pursuing the policy of monopolistic globalization through the machination of WTO. I nursed no great hope that something positive will emerge out of my interactive session with Mr. Pascal Lamy, Director General, WTO, who visited New Delhi on 5th April ostensibly for having a better appreciation and understanding of India's over all approach and implications of the on-going WTO negotiations.

For the stakeholders, expectations never ran high and there was no misgiving that the visit of Mr. Pascal Lamy to India was aimed more at scoring even a better deal for the developed countries than appreciating the problems that throttle the livelihood of the poor and vulnerable segment of the third world, the agro-rural people.

He has never been effective in reducing or phasing out of farm subsidies by the developed countries. Absolutely imperative as it was, my dialogue with Mr. Pascel Lamy was an extension of our effort to safeguard the interest of developing countries. During an interactive session with him on 5th April 2006 in New Delhi, it was made absolutely clear that dilatory tactics and threats might ultimately cause undoing of the WTO itself. While taking over from Mr. Supachi Panitchpakdi of Thailand, Mr. Lamy committed himself to fair and justifiable trade negotiations and work out solutions on contentious issues of subsidies, domestic support, market access and various other distortions in AoA. But as Director General of WTO, he has been consistently downplaying the concerns of developing countries.

WTO's core motive is to promote the interest of agri-business multinationals at the expense of small and marginal farmers and family farms across the world. Agriculture negotiations should be from the point of view of the farmers and not from the point of view of multinationals and economists. There are sufficient reasons to take a critical view of the present subsidy regime as large chunks of subsidies and support given in North America and Europe go to agri-business corporations and farmers with large land holdings. All trade experts agree to the fact that subsidies and support are trade distorting as they depress global prices placing the farmers in the developing countries at a disadvantage.

Notwithstanding the fact that all subsidies are trade distorting, some so-called trade experts and negotiators of the developed world try to justify their misdeeds by categorizing subsidies and support as "trade-distorting" and "non trade-distorting". Even WTO sings to this tune of categorization. Despite their commitments to reduce subsidies, the developed countries have increased their subsidies to escalated levels. Agri-business corporations and large farmers are strong economic entities and therefore should not need any support.

The present subsidy regime is contrary to the general perceptions that the strong need no support. Farmers do not agree with the jugglery of words in WTO literature like creation of different Boxes-Green, Blue, Amber etc.- and different types of sugar coated formulae. If the mission of Mr. Pascal Lamy was for ensuring free and fair trade he should take urgent steps to phase out all types of subsidies and support to agri-business corporations and farmers with large land holdings in the developed countries. Export subsidies and support are meant for traders and should be phased out immediately.

Small and marginal farmers, particularly in the Third World, need to be protected against cheap subsidized imports. The concept of special products (SPs) and special safeguard mechanism (SSM), now being floated, is inadequate to safeguard the livelihood of farmers. These mechanisms will invite unnecessary complications. In lieu of these new concepts, it is imperative that developing countries be given the option to apply quantitative restrictions (QRs) on imports, whenever needed, to protect the livelihood of farmers.

Agriculture is not only for trade, it's a way of life in the developing countries. European Union accepts this and says that agriculture is multi-functional. It is amazing why European Union should not come out openly and say that it is high time to bring back the mechanism of QRs on imports to protect the farm sector. Interestingly, WTO has now come to decide what type of food should be consumed by the people who may be against the national and cultural habits and prove to be hazardous to health and environment.

The WTO dispute settlement body has recently given a ruling against European de facto moratorium on GM foods. This ruling amounts to forcing countries to accept food much against their choice. WTO should not dictate the type of food that should be consumed by consumers in different countries.

The WTO should focus mainly on the concerns of the producers of primary commodities, the farmers and not for the interests of agri-business multinationals.

Source : Presentation in the United Nations Conference on Trade and Development and Ministry of Commerce (Govt. of India) meeting; April 5, 2006, New Delhi

45

Video Conferencing on 'Doha Development Agenda' Organized by the World Bank

Agriculture in India is not just an occupation or business but for the people it is the way of life, very much enshrined in the socio-economic fabric of the country. The developing countries should safeguard the interest of their large number of small and marginal farmers, ensure food security for the people and provide livelihood for their agricultural workers.

Farmers in developing countries have been severely hit by domestic support and export subsidies provided by developed countries to their farm sector. This has depressed global prices (distorted prices), leading to India becoming a dumping ground of heavily subsidized agricultural products.

Developed countries have not only blocked market access for developing countries but are trying to push their heavily subsidized farm produces and commodities into the markets of developing countries.

The developed countries have not fulfilled their commitments to reduction of subsidies and supports stipulated in the first phase as per the agreement on agriculture (AoA). Instead, they have increased subsidies by 50% through shifting it from one box to another. It has become very difficult for developing countries to compete in the world market. It totally goes against the spirit and provisions of AoA.

It has become evident that developed countries, which are pumping in massive farm subsidies to the tune of $ 1 billion a day, would try to continue providing support to their farmers in one way or the other. The US Senate rejected a 30 per cent cut in the annual farm subsidy payments.

It will further aggravate the trade imbalance between the developed and the developing countries. India shall have to be very firm that unless there are real reductions in subsidies, the question of lowering its bound tariffs does not arise. Remarkably, the US exports to India has increased substantially.

The US investments in India are aimed mainly on technologies. India being one of the largest farming countries in the world and agriculture as the major livelihood resource of the people, it becomes highly imperative to adopt a very clear and farmer centric strategy in all agriculture related negotiations.

The objective should be to eliminate all subsidies and supports by all developed countries with immediate effect. The developed countries have so far not fulfilled their commitments on Agreement on Agriculture within the prescribed time frame. All subsidies are trade distorting; they create serious trade imbalance between developed and developing countries.

The developing countries should never agree to reduction of tariff unless total subsidies and supports are phased out by the developed countries. The situation being so, question of market access does not arise unless the subsidies and supports are phased out. Both additional tariff and quantitative import restrictions should be admissible as special safeguard measures for developing countries.

The developing countries should avail special safeguard mechanism (SSM) for all products and must also have the right to select and limit special products of their choice. The issue of sensitive products of developed countries should be a matter of intense negotiations. The peace clause should not be renewed in any form.

WTO has adversely impacted Indian agriculture and its farmers stances as under are self explanatory and indicator of the real situation:

The MSP for paddy is Rs. 530/- per quintal while in Orissa, Bihar, and Madhya Pradesh the growers are forced to sell paddy from Rs. 425/- to Rs. 450/- per quintal.

The MSP for Mustard is Rs. 1715/- per quintal. In Rajasthan the growers were compelled to sell Mustard from Rs. 1425/- to Rs. 1450/- quintal, much less than the MSP.

Source : Kisan Ki Awaaz; Nov 30, 2005;[http://www.kisankiawaaz.org/Video%20 Conferencing%20on%20%E2%80%98Doha%20Development%20Agenda%E2%80%99 Organised%20by%20the%20World%20Bank%20on%2030th%20Nov.%202005.html]

46

GATT IT? Got it

Bureaucrats who are supposed to guard India's interest at international negotiations have repeatedly let the country down. Krishan Bir Chaudhary reveals how senior officials have compromised on national interests at international forums and got rewarded by the powers-that-be

"India could cooperate with developed countries on agriculture if they cooperate with India on Non Agriculture Market Access (NAMA) and services." - Union Commerce minister Kamal Nath on 2nd May 2005 at the OECD and G-20 meet in Paris.

FROM DOHA 1995 TO Geneva 2004, the 10-year-long treacherous journey for India as a founder-member of WTO has proved all enduring and elusive, specially for its agro-rural sector. The Uruguay round of negotiations preceding Doha declaration has been a sordid saga of bureaucratic muddle by the Indian negotiators.

The government was misled into signing the Doha Declaration on the dotted lines, compromising national interest that landed India inextricably trapped into the sinister machination of WTO, having the back up of super economic giants.

Counter positioning India's strategy and economic interest must be termed as the worst form of bureaucratic impropriety which can never be condoned.

It is not a strange coincidence but a reality that the more they compromised the national interest, the more they were awarded. Otherwise, how can one explain that Mr. Anwar Hoda, who was special secretary in the ministry of commerce in 1988, and chief negotiator for India during the Uruguay round, got the plum post of deputy director general, GATT and right now is serving as member of Planning Commission.

Mr A.N. Verma, secretary in the ministry of commerce led the delegation to Geneva in April 1989. Taking advantage of the then prevailing political situation, Mr. Verma told Mr. Montek Singh Ahluwalia, economic advisor to the Prime Minister, that since dispute settlement mechanism in the new agenda for the Uruguay had been strengthened, therefore, it should be accepted.

Accordingly, as advised by Mr. Ahluwalia, the Prime Minister agreed to accept the new agenda which included IPR, services, investment and domestic policy related to agriculture. Mr. Verma was later promoted to the rank of principal secretary to the Prime Minister.

Mr. A V Ganeshan who was secretary, commerce, in December 1993 during the final act of Uruguay round, toe-toed the line of his predecessor and after retirement he was rewarded as the panel member of the dispute settlement body.

Mr. Narendra Sawarwal was joint secretary (patent) in the industry department in the early 1990. A bureaucrat belonging to the same bunch of thought, he was rewarded with the position of director of Asia Pacific Bureau of World Intellectual Property Organisation (WIPO). At the moment, he is enjoying as its senior director.

Mr. Pushpendra Rai, now working as the director of WIPO, was a joint secretary (patent) in the industry department in the mid 1990. Likewise, Dr. Jaiyia served as director in the ministry of industries dealing with patent matters. Now he is enjoying his stint in WIPO.

Mrs. Jaishree Watal was a deputy secretary with the ministry of commerce during the Uruguay round negotiations. Now she is enjoying her assignation in the WTO.

Media exposure of Mr. A.E. Ahmed, joint secretary industry department (patent) who did a U turn and compromised government's actual position to support the WIPO director general's patent agenda during the WIPO assembly and his questionable role in the patent amendment bill - 2nd and 3rd amendment failed him from being rewarded like others.

Before the Uruguay round, only two issues - trades and goods - were in the agenda for bilateral or multilateral negotiations. But it was the collaborative venture of these gentlemen which brought various other non-conventional matters in the Uruguay round agenda.

India and Brazil led the third world in Uruguay round negotiations during 1986-96. We have seen the angry outcry and protest against the outrageous and unequal globalisation through the machination of the WTO.

The stiff resistance put against the Doha declaration in Montreal by people from across the world will go down in the annals of our times as the largest human globalisation of the sort.

We call upon the government to table a white paper in Parliament pertaining to WTO negotiations.

The government should also put stringent moratorium for at least five years banning all bureaucrats from joining any pecuniary position anywhere who are involved in the WTO negotiations.

Source : COMBAT LAW magazine; June-July, 2005;[http://www.combatlaw.org/?p=848]

Dangers of Genetically Modified crops & MNC's

47

Biotechnology Regulatory Authority of India Bill in Parliament

The Biotechnology Regulatory Authority of India (BRAI) Bill is to be introduced in the Parliament. The ostensible stand of scientists defending genetic engineering technology in seed production is that utmost priority to a product's impact on humans, plants, animals and environment shall be given.

The first problem arises out of the claim that "scientists are experts in their respective areas of specialization." This is farthest from truth. It is a known fact nearly all Agriculture biotechnology scientists have sold their souls to the seeds multinational corporations and work for them; only a handful remain "independent" and increasingly under pressure to fall in line.

In fact, how systematically just a handful of powerful seeds corporations destroyed scientific integrity of world's leading scientific establishments itself serves as hard evidence of what the present "scientific establishments" have become.

The second problem arises out of lack of transparency. The Bill was earlier prepared under Official Secrets Act. Why? Was national security at stake? In fact, yes. The Bill undermines national security, it undermines food security and food sovereignty, and it seeks to compromise public health. Since the Bill seeks to destroy three vital aspects of India as a nation.

India is the last bastion of food mega-diversity and seed mega-diversity. Until and unless the seed system and food system come under control of just five multinational corporations based in the USA and EU, they are not going to rest.

The third problem is that the Bill seeks to give primacy to genetically engineered seeds. Why should company manufactured seeds have primacy over farmer saved and farmer bred seeds? Is corporate profiteering more important that people health and India's food security?

The fourth problem is the silence on "polluter pays" principle. Thus far the polluters have not paid, nor compensated any farmer for destroying their natural or organic farms anywhere in the world.

Only when farmers lodged court cases that in some instances some have been compensated but not to the extent they should have been. The tide is slowly turning in some countries and we in India are not far behind.

The producers of conventional and genetically-modified (GM) crops and the users of proprietary pesticides and herbicides that accompany them are now facing serious legal battles from natural and organic farmers.

All four issues raised here stem from basic human rights. If the Bill does not address even one of these, it should be scrapped.

Source : Editorial; Kisan Ki Awaaz; August, 2011;[http://www.kisankiawaaz.org/Biotechnology%20Regulatory%20Authority%20of%20India%20Bill%20in%20Parliament.html]

48

Data Protection will Benefit MNCs

On behalf of millions of small and marginal farmers of India and in continuation of our letter dated 16th July 2010 and the reply of that letter given by the Agriculture Ministry, I am writing the facts related to the Pesticide Management Bill-2008.

Regarding protection given in other part of world, why Agriculture Department comparing our developing country with the developed countries. When the Patent Law already exists in the Country then what is the need for double benefit (extra time) is being given to MNC's for monopoly of their so called new pesticides, most of the molecules are obsolete in many Countries.

More than five hundred pesticide molecules are waiting for entry into Indian markets. MNC's will sell such products at exorbitant prices to the Indian farmers. They are trying to capture the input sector of agriculture.

We should not give the monopoly rights to MNCs unnecessarily by giving the tool of data protection. We have raised all issues before the Parliamentary Standing Committee on Agriculture also because it's our moral duty to protect our poor farmers and we should not surrender the input sector of agriculture to the multinational corporations. Why is the government adopting double standard with the most important input sector of agriculture, which is directly related with the food security of the country? This act of the government will increase the cost of production.

When there is no data protection in the pharmaceutical industry in India, which is also equally important sector, then why in the agriculture sector? Once the data is generated in any part of the world and the molecule is found to be efficient on crop and safe for the environmental, animal and human health then further registration of the same molecule from other source is unnecessary and sheer wastage of money and energy. This will prevent Indian industry to give the same molecule to the farmers at cheaper price.

Any new molecule invented in the world for the first time with commercialization in any country all over the world is already patented then why should we allow a separate data protection when 17 years' protection is already guaranteed.

Since India is already signatory to WTO on Patent/IPR and this covers 17 years' protection of data and it is WTO the sole authority who can increase and decrease the period of patent then why are we unnecessarily granting further 5 year data protection. Over and above this is prerogative of international agency i.e. WTO or the patent authority to increase or decrease the period of data protection.

The data protection will be misused by misleading the government and making the interpretation by MNC's in their own interest through our corrupt government officials.

Every time they twist their data in the name of new data and ask for further protection for 5 years, they would be using this protection for ever. It shows how the powerful MNC's are influencing our policy makers and pressuring the government.

Source : Editorial; Kisan Ki Awaaz; Oct, 2010;[http://www.kisankiawaaz.org/Bharatiya%20Krishak%20Samaj%20letter%20to%20Government.html]

49

Government should place Moratorium on all GM Crops

Bharatiya Krishak Samaj is against the irreversible, uncontrolled, and potentially dangerous release of genetically modified organisms (GMOs) into the environment. We are also concerned about the health hazards of GM products. The International Scientists community have given their reports and proved that GM Crops cannot increase the productivity and cannot feed the hunger.

There should be Mandatory Labelling of all GM products, irrespective of the percentage traces of GM materials present in the product in any form whatsoever. Labelling norms should clearly state that it is a GM product. There should be no soft provision to state "May Contain GM Traces." GM products is entering in the market illegally through imports. As of now only approved GM product in the country is Bt cotton.

But we are sorry to note free clandestine imports of GM seeds and GM food and feed into the country, flouting all regulatory norms. Unapproved imported GM seeds are being cultivated in this country, in blatant violation of bio safety norms. Adequate tests should be done in the country to establish the safety of GM products to be approved for imports. Test should establish safety of health and environment. The process and results of such tests should be made transparent. If such adequate tests cannot be conducted in the country to establish health and environment safety, precautionary principles of the Cartagena Protocol on Bio-safety should be applied to deny imports of GM products for food, feed, processing and cultivation.

However for research purposes, imports may be allowed, but with a strict provision of not allowing its release in the environment. Caution should be taken that the research materials are not clandestinely diverted for use in food, feed and processing for consumption.

Here is the relevant portion of the Cartagena Protocol: "lack of scientific certainty due to insufficient knowledge regarding the extent of the potential adverse effects of a modified organism on conservation and sustainable use of biological diversity in the Party of import, taking also into account risks to human health, shall not prevent that Party from taking a decision, as appropriate, with regard to the import of the living modified organisms in question.....to avoid or minimise such potential adverse effects."

There are in pipeline a number of GM food crops for approval. There is no sufficient sophisticated mechanism in this country to establish the safety of GM crops. The most worrying situation has been created by the Indo-US Knowledge Initiative in Agriculture Research and Education. Through this pact US is putting pressure to relax our regulatory norms on GM products.

Now the government is planning to dismantle GEAC of environment ministry and to set up an autonomous regulator under ministry of science and technology for all GM products. Such a move would be dangerous. The GEAC should continue under Ministry of Environment and issues related to environmental, human and animal health should be vigorously addressed with transparency. Health Ministry and ICMR should be more responsible and they should be aggressive in expressing concerns over the safety of GM foods.

Source : Editorial; Kisan Ki Awaaz; May, 2010;[http://www.kisankiawaaz.org/Govt.%20Should%20put%20Moratorium%20on%20all%20GM%20Crops.html]

50

Enforce Strict Liability Regime
to Address Damages Caused by GMO's

It is high time that India live up to the principles of the Cartagena Protocol which it has already ratified. The health and environmental hazards of genetically modified organisms (GMOs) as revealed by leading independent scientists are matters of grave concern.

The world leaders have admitted the situation and have adopted a global treaty called the Cartagena Protocol to address the problem. This treaty has come into force since September 11, 2003. The Article 27 of the Cartagena Protocol calls for setting up of a global liability and redressal mechanism for damages caused on account of trans-boundary movement of GMOs otherwise called living modified organisms (LMOs) in the treaty.

The Protocol also calls upon countries for adopting precautionary principles for addressing the likely threats of GMOs. Setting up of a global liability regime and redressal mechanism for damages caused on account of trans-boundary movement of GMOs is currently under discussion. India as a party to the Protocol has called for a stringent global liability regime.

India, which is assuming leadership in several global fora on many critical issues, needs to set examples at home. It should not only adopt precautionary principles to meet the threats of GMOs but also adopt a stringent liability regime to address the damages caused on account of GMOs.

So far India has approved only one GM crop Bt Cotton for commercial cultivation. Cultivation of Bt cotton has invited a host of problems. The assurances given by Monsanto and Mahyco have failed and farmers are facing heavy losses on account of cultivation of Bt. Cotton. The incidences of suicides among farmers have increased since the introduction of Bt cotton.

Bt Cotton has attracted new pests like mealy bugs. As bollworms and insects became resistant to the single transgenic Bt Cotton, Monsanto and Mahyco introduced stacked genes Bt Cotton and even this new product has failed to give the desired results.

Sheep and goats grazing over Bt Cotton fields faced death. This proves the health hazards and poisonous effects of Bt genes. The Genetic Engineering Approval Committee (GEAC) instead of expressing concerns over these incidents are pushing for approval several GM food crops in the pipeline at the instance of seed multinationals like Monsanto.

It has given approval to Bt Brinjal without conducting adequate bio-safety tests. The government needs to wake up and enforce precautionary principles to meet the threats of GMOs and put in place a stringent liability regime to book the culprits like Monsanto and other GM seed companies to pay for the damages caused to the farmers and the nation.

We should learn from the Bhopal Gas tragedy how severe the consequences can be if we are not vigilant and do not adopt any precaution. Bt Cotton has already caused damages. Let us wake up and prevent the damages likely to be caused by other GM crops.

Source : Editorial; Kisan Ki Awaaz; April, 2010;[http://www.kisankiawaaz.org/ Enforce%20Strict%20Liability%20Regime%20To%20Address%20Damages%20Caused %20by%20GMOs.html]

51

Bt Brinjal Fraud Exposed

The year 2010 began with a new note in India which may be remembered in the history as a turning point where the people for the first time came out in openly questioning the scientific fraud of illegal approval of genetically modified food crop, Bt Brinjal.

After the Genetic Engineering Approval Committee (GEAC) approved the Bt Brinjal developed by Mahyco in collaboration with the US seed multinational Monsanto on October 14, 2009, the Union minister of state for environment and forests, Jairam Ramesh took the wise decision to withheld its release and go for public consultations.

The advocates of the GM technology thought they would win the race by tackling a handful of NGOs. But this did not happen. People on their own came out in the open opposing the introduction of Bt Brinjal be it in Kolkata, Bhubaneswar, Nagpur, Ahamedabad, Hyderabad, Chandigarh, Bangalore. The protest turned violent in Kolkata.

Farmers too joined in the protests. The people have come to know the fraud committed by a handful of scientists in pushing the approval of Bt Brinjal.

A noted molecular biologist and a special invitee to the GEAC at the instance of the Supreme Court, Dr Pushpa M Bhargava has pointed out the flaws in the approval process and accused the regulator for skipping adequate bio-safety tests.

Admittedly Bt is a toxin. Sheep and goats grazing over Bt cotton died and GEAC did not take any cognizance. Independent scientists across the globe have conducted several studies exposing the health and ecological hazards of GM crops.

Now the Indian people have come to know that there are two classes of agriculture scientists one class who on the payrolls of seed multinationals like Monsanto and the others who are pursing science in public interest.

The objective of the seed multinationals is clear. They want to control the entire food chain. The Monsanto charges heavy technology fees. The entire game plan is backed by the US which is engaged in the geopolitics of capturing global food market through the introduction of GM crops.

Public protest against Bt Brinjal has drawn the attention of the political parties ranging from the right wing parties to the left wing parties.

Source : COMBAT LAW Magazine; March 25, 2010;[http://www.combatlaw.org /?p=227]

52

Do not force GM Food through Draconian Law, go for Public Referendum

A new game plan is in the offing after the denial of the release of the Bt Brinjal for commercial cultivation. The Union ministry for science and technology is gearing up to table the National Biotechnology Regulatory Authority Bill in the Parliament.

The science and technology ministry and its department of biotechnology are aggressive promoters of genetically modified (GM) crops, unmindful of its health and environmental hazards revealed by a series of studies done by scientists across the globe.

The Union environment and forests minister, Jairam Ramesh has done the right job. After the Genetic Engineering Approval Committee (GEAC) gave its green signal to Bt Brinjal he exercised his discretion to withhold its release for commercial cultivation.

The GEAC is after all under the administrative control of the environment and forests ministry. The environment minister opted to go for public consultations across the country. The result was that the majority of the people voiced their opinion against the release of Bt Brinjal.

The GEAC had not its homework properly before giving its nod for the clearance of Bt Brinjal. Several lacunae was pointed out by the eminent molecular biologist, Dr. Pushpa M Bhargava in the GEAC meetings to which he was a special invitee at the instance of the Supreme Court. But GEAC dominated by persons having vested interests ignored Dr Bhargava's suggestions and biosafety norms and pushed for the release of Bt Brinjal. The lobbyists for GM crops have now found a new friend in the Union science and technology minister, Prithviraj Chavan and have urged him to table a draconian law in the form of the National Biotechnology Regulatory Authority Bill so that the voices of public resentment will be silenced.

The Union Food and Agriculture Minister, Sharad Pawar has openly come out favouring the release of Bt Brinjal. According to a leaked out news the most obnoxious part of the proposed draconian law is to impose penalties if there is any criticism of modern biotechnology by any citizen be it a writer, journalist, scientist, research institution or university.

The penalties for "misleading public about organism and products" are 6 to 12 months of imprisonment and fine up to Rs 2 lakhs. The decision of the proposed authority will be out of the purview of law courts. Such a draconian law would not be keeping with the traditions of a vibrant democracy. In the colonial regime India did not had such a draconian law where the state would dictate the people to consume hazardous food.

India, being the largest democracy, should set examples and enrich democratic tradition. Hazardous GM food should not be imposed on unwilling people through draconian laws. It would be wise for the government to take a countrywide referendum on GM crops, instead of relying on the suggestions of handpicked scientists who act on the directions of multinational GM crop industry.

In a democracy people should have the right to choose what they should eat not the government or seed multinationals.

Source : Editorial; Kisan Ki Awaaz; March, 2010;[http://www.kisankiawaaz.org/ Do%20not%20force%20GM%20food%20through%20draconian%20law,%20go%20for %20public%20referendum.html]

53

On the Proposed Field Trials of Bt Brinjal

This is with regard to the Agenda Item No. 4.2 in the GEAC meeting dated 1/6/2006 which says that GEAC will consider permission for seed production and Large Scale Trials of Bt. Brinjal of four Mahyco hybrids in this meeting.

In this regard, we would like to bring to your notice that the GEAC is yet to act and fix liability on the various biosafety violations and irregularities brought to your notice by civil society organizations, including on a Bt Brinjal limited field trial in Andhra Pradesh.

While turning a blind eye to the harsh realities related to serious regulatory failures, you seem to be ready to go all out to support the industry in its profit-making endeavours.

Given that there are many studies on adverse health effects with many GM crops from all over the world, what is the assurance that the right questions have been asked in this case to arrive at the right answers with regard to the safety of these crops? We have found that Bt Cotton, which was upheld through your tests to be safe to human health is indeed causing a lot of health problems amongst farm workers and ginning factory workers.

Similarly, there were recent reports on the mortality of livestock after grazing on Bt Cotton. It is not clear how GEAC is assessing such possibilities as part of its bio-safety testing regime nor how is it acting at least on such reports by commissioning detailed independent investigations once these preliminary reports are out.

It would be disastrous for this country to rush into approvals just because a prescribed set of tests for a prescribed period have been completed as per some procedures laid down.

There has never been a serious public debate initiated on the adequacy of such a bio-safety regime and even though there is a Supreme Court case pending precisely on this matter, the GEAC seems to be in a great hurry to approve even GM food crops like vegetable crops, for reasons that are not clear at all.

This is especially surprising in all those crops where safer, inexpensive and farmer-controlled options like IPM, NPM and Organic are in place, successfully practiced and established all over the country.

The data presented by the company on various studies done in Bt Brinjal was put up by the GEAC only this morning [31/5/06] and considering that this is a very important food/vegetable crop of the country, there should be at least 90 days allowed for feedback on the bio-safety tests and their findings.

Secondly, the data put up is not adequate for an intelligent and scientific debate to take place since it only has findings without details of the research design and protocol in each case. We demand that full reports on each test be presented and not just findings.

We also demand that the GEAC show its accountability to the public by sharing what improvements have been made in its bio-safety protocols, in its monitoring systems and in its accountability systems, before giving any more permission for trials, given your proven inability to ensure bio-safety in this country.

Here, we would like to remind that it was during such field trials that illegal Bt Cotton and rapid contamination of the Cotton chain began in this country and GEAC could only wring its hands in helplessness.

The dangerous and unscientific manner in which field trials take place in this country tell us that we are only one step away from a huge bio-disaster wreaked on Indian agriculture.

To sum up, we once again demand that the complete set of data on Bt Brinjal including the testing protocol be put up for public feedback, that at least 90 days be provided for such feedback, that such feedback be taken on board with all the seriousness it deserves, that clearly-needed improvements be made in the testing regime and monitoring systems and shown to be made to the public before any permission for any more trials in the open environment are given in the country.

Source : www.gmwatch.org ; May 31, 2006;[http://gmwatch.eu/latest-listing/45-2006/5272-stop-bt-brinjal-1162006]

54

Transgenics and Indian Agriculture: where are the benefits?

In India, transgenic crops are being experimented with and even released, without a coherent approach to the whole matter. It is not clear why transgenic agriculture is considered "frontier" or indispensable by numerous agricultural research bodies both in the public and private sectors. Given below are our strong objections to transgenic crops in Indian agriculture and the reasons for the same. Firstly, when it comes to transgenic agriculture, it is not clear how research and commercial release priorities are being set in this country. It seems that agencies are free to choose what suits and benefits them most, rather than what farmers need. No consultation with farmers and their organizations on whether they want GE as a technology in Indian agriculture at all is visible.

There is no assessment witnessed of various options - including safer, more inexpensive and politically right decisions that would uphold farmers' interests - before zeroing in on transgenic technology as the option for a given crop or problem. Herbicide resistance is a trait that is being worked on by many agencies, including public sector bodies! What implications would this have for the poor agricultural workers of this country, not to mention the environmental implications with increased herbicide use? Similarly, major food crops are being worked on without any thought to environmental and health repercussions! This includes our staple food, Rice. Public monies are being spent on expensive research on crops like tobacco! How are these research priorities being set? What are the accountability systems here, given that public sector research is much more than private sector when it comes to transgenic crop experimentation in India?

Also, how have agencies, especially in the public sector, zeroed in on research on transgenic, rather than research on safer, ecological alternatives? These public sector bodies shy away from even validating such ecological practices that are being adopted by farmers on the ground. They would rather spend expensive resources sitting in their laboratories and campuses developing an imprecise technology. On top of this are complications related to IPRs which have not been worked out at all. The UAS, Dharwad has a case to illustrate where they had developed a Bt Cotton variety with a gene donated by Ford Foundation only to discover later that that gene is a proprietary technology owned by Monsanto!

Civil society verification and research shows several bio-safety violations in all such experiments - the products from field trials are allowed to enter the food chain routinely before all bio-safety tests are completed. Seeds from such field trials are routinely allowed to contaminate the other seed stock either physically or biologically, much before such crops are allowed for commercial cultivation. The Navbharat Bt Cotton fiasco would have happened in such a manner too though no detailed investigations were undertaken on the matter. Similarly, field trial permissions and seed production permissions are given and no monitoring takes place to check what happens to the seed stocks if commercial approval is not granted in the next season.

There have also been instances in the past where attempts have been made for clandestine imports of GE foods into the country or when they have actually been imported. The Soya imports into this country from countries like the US must surely be GM-contaminated - however, no permission for such imports is being sought from the GEAC nor is GEAC pro-actively stopping such imports. All of these are clear indications of the complete failure of bio-safety regulations or risk assessment procedures in the country.

Coming to the experience of Bt Cotton in India, the first transgenic crop to be commercially cultivated, there are many lessons to be learnt including the fact that the technology is very imprecise and imperfect. Government's own studies have shown that Bt Cotton, a technology imported from the US, was fit for the American conditions and their major pests rather than ours. There has been an extremely uneven performance, predictably, of the technology on the ground - the primary claims have been belied with regard to pesticide use coming down along with bollworm incidence coming down. There are several other problems reported by farmers which need deeper investigations - this is however not being done despite repeated requests.

The country has not stopped to pause to take stock of the situation so far, before more varieties are released all the time. Worse, bio-safety assessments are being done away with, with the argument that the "event" has already been approved for its bio-safety. This is a highly questionable claim. Bt Cotton cultivation in this country has also shown all the shortcomings and lacunae in our regulatory functioning. The post-approval surveillance is extremely unscientific and erratic. The cases of falsification of actual experience on the ground point to corrupt elements entering the picture. Monsanto's bribing of several Indonesian officials for obtaining a clearance for a GE crop is well-known and is a good reminder to us about the extent the industry would go to push its markets.

The most important shortcoming in the story of Bt Cotton in India has been the lack of accountability mechanisms. Farmers who have incurred losses due to the cultivation of Bt Cotton have been left to fend for themselves while the companies involved in the commercialization are laughing all the way to the banks. Farmers' interests have definitely been shown to be the last priority in this fiasco. Resistance management plans are non-existent and faulty where they exist. Even in a country like Australia, there is a 30% limit to Bt Cotton cultivation. Why do Indian scientists only talk about experiences from elsewhere and adverse results from their own studies, instead of doing something to influence the decisions? Is scientific research by specialist bodies like CICR meant only for academic interest? India should also take cue from the developments across the world. Worldwide, starting from 2003, GM crops research is drying up, even in countries like the US. Companies like Bayer Crop Science have announced that they are going back to conventional breeding. Companies are also voluntarily withdrawing products that have been in the pipeline like GM Wheat due to enormous consumer and farmer pressure against these crops.

India should consider why it wants to tread a path that could be inimical to the interests of its farmers and definitely prove hazardous to its environment. Let us look at the situation worldwide - In 2004, the biotech industry and their allies celebrated the ninth consecutive year of expansion of genetically modified (GM) crops. The estimated global area of approved GM crops was 81 million hectares in 22 countries. Corn and soya, the two most widely grown GE crops are grown mostly for animal feed or enter the human food chain mostly as minor ingredients or derivatives. The GM industry would like to tell us that it has delivered benefits to consumers and society at large through more affordable food, feed and fibre with less pesticide usage.

It is difficult to imagine how such benefits have been achieved given that more than 70% of the global area under GM crops is devoted to Monsanto's Roundup Ready herbicide-tolerant crops. Even yield increase claims are questionable since studies from the US show that yields were suppressed with crops like RR Soybean cultivars. Other studies from North America on Roundup Ready Soy and Bt Maize found that the returns from these crops essentially equalled those of non-GE varieties.

The social costs of displacement of small farmers and agricultural workers from their farming are well documented and enormous. In Argentina, the situation is quite dramatic as 60,000 farms went out of business while the area of Roundup Ready Soybean almost tripled. In countries like Brazil, GM soybean-led deforestation of the Amazon forests is also well- documented.

These developments only point towards the very hollow impact assessment studies and risk assessment studies that are taken up before the introduction of the technology. Often, such studies are not taken up at all and India cannot be allowed to go the same way. Please note that there are strong reasons as to why only 22 countries in the world have so far approved GM crop cultivation. The environmental costs of the transgenic technology in agriculture are irreversible and unaffordable. Degradation of soils, loss of sustainable farming practices, loss of biodiversity, huge monocultures to the detriment of the sustainability of resources, impact on other living organisms, increase in secondary pests' damage to the crops etc. have all been well-documented. Equally well-documented are the positive impacts of many sustainable agriculture practices which are non-pesticidal and non-GE.

The use of chemicals has only increased after the introduction of GE-led agriculture in countries like the US. In 2004, farmers sprayed on average 4.7% more pesticides on GE crops than on identical conventional crops. In the case of herbicide resistant crops, the usage of herbicide goes up and in the case of insect-resistance crops, insects are known to adapt themselves given the enormous selection pressure on them which once again translates itself into higher chemical use for their control. The increase in chemical usage not only has environmental implications in terms of groundwater contamination, super-weeds etc., but also raises important questions on food safety.

Coming to the much-forwarded principle of co-existence of GM and conventional crops, regulators and scientists should understand that co-existence is impossible in India. Experiences from world over including the Mexican maize contamination case are an illustration. "Adventitious presence" or contamination of conventional seed with biotechnology traits is a known phenomenon which has adverse environmental and economic implications.

In a country where there are millions of small holdings right next to each other and where traditional seed exchange systems are vibrant to this day, both genetic and physical contamination of seed stocks is inevitable. Failure of regulation is more than well-established in the case of a non-food crop like Cotton. The disaster waiting to happen if GM technology is introduced in food crops cannot be overstated.

GM foods are known to cause a variety of human health problems. There are numerous studies on GM tomato, GM potato, GM corn, GM soy and other crops which show that these foods constitute a definite hazard to health. Monsanto's secret GM Maize study findings also point to the same facts. There is also the issue of antibiotic resistance building up through GM crops. There can be no easy management solutions to these issues. In developed countries too, segregation was known to have failed as the Star link corn contamination case reveals.

Many long term human health impacts might not even start showing in the health assessment studies being taken up right now. How can India afford to tread this path, when it has agreed to enshrine the Precautionary Principle when it signed up to the Cartagena Protocol? How can the precautionary principle guide us for international trade decisions but not when it comes to domestic production and trade decisions?

Has India begun assessing the possibilities of market rejection for its agricultural products if it opts for GE any further? Many large companies in the mainstream food industry already have a non-GE policy in response to consumer demand in many countries in the West. What will be the economic implications for Indian farmers of such market rejection? What kind of an analysis is available for the farmers so that they can make an informed choice on the matter?

The organic food industry, which has a great potential for growth will definitely be closed to us by our pro-GE decisions and this will once again mean a great economic loss to Indian farmers. Organic farmers have their own rights which need to be protected too. In Canada, a class action suit is under way demanding lost organic canola profits due to contamination. Similarly, Germany has a law that makes farmers who plant GE crops liable for contamination of other crops.

Many other countries in Asia are treading cautiously and have moratoriums, or bans, or pro-active organic farming policies in addition to strict labelling regimes for regulation of their agriculture and food industry. India however seems to be moving in a very ad-hoc and anti-farmer manner in this regard.

India often talks about emulating the USA without considering that the social and agro-ecological conditions are vastly different between the countries, not to mention the regulatory mechanisms. India has to evolve solutions for its agriculture indigenously and an enormous number of successful alternatives to various situations exist with the farmers themselves in various pockets of the country.

It is time that the agricultural research establishment, the agricultural education establishment as well the agricultural policy-makers first look at options before chasing technologies that are unsustainable and anti-farmer.

Source : The paper was presented in National Commission on Farmers September 22 ,2005; [http://www.gene.ch/genet/2005/Nov/msg00013.html]

Some Glimpses
of Author's Activities

55

FDI in Retail Sector will not Augur well for India

Belgaum - May, 5 - Bharatiya Krishak Samaj (BKS) National President Krishan Bir Chaudhary here on Friday asserted that multi-national companies (MNCs) will capture food security of the country through Foreign Direct In Vestment (FDI) in Retail.

Addressing a press conference here he said there were many examples in the developed countries that FDI did not benefit the farmers. The rural India is facing an agrarian crisis and 68 per cent population of the country is depended on agriculture. The 85 per cent farmers are small and marginal and having less than 2 acre of land. In India the retail sector give employment to 40 million people. We cannot compare with China because India's priority is national interest and poor people's welfare. When the government cannot provide the sufficient employment to the people then why the government is destroying the self employed sector of the country. He asked.

Mr. Chaudhary said for MNCs it was agri-business but for us it is way of life. The FDI in Retail is nothing but a game plan to control the food security of the country through the contract farming system. We will lose right on our natural resources i.e. land, seeds, water and environment. After their monopoly the farmers had to dance to the tunes of these MNCs. Independent farming and trading were the pillars of strong India. Why the government was not ready to develop the infrastructure in the country he questioned. The agriculture need better marketing better infrastructure and effective network in the country. The traditional retail culture cannot be replaced by the MNCs system he claimed.

He said if the food security of the country was controlled by these MNCs through their patented GM seeds and inputs in agriculture sector the farmers would be exploited. The economy of the country will be controlled by them because food is the main currency of India. If FDI in multi brand retail was allowed in the country then our farmers will suffer the same fate as their Western counterparts he opined.

Replying to a question Mr. Chaudhary said the argument that inflation will be reduced by reducing middlemen is false because they just replace small traders with a giant trader who collected high profits. The hyper markets displace diversity quality and taste. Their giant retail companies will not just destroy our retail and farmers but also devastate our culture and very social fabric.

He said the huge subsidies for farmers in developed nations showed that FDI did not benefit farmers are perishing over the globe. As far as the consumers are concerned they will be negatively affected too because as seen from the experience in other countries like Mexico Argentina and others the supermarkets usually sell more expensive food than other small and informal outlets. The protests in Arab countries and other parts of the world amply reflect the destruction of small retail stores which has resulted in mass unemployment he added.

To another question he said BKS proposed to form a federation of all farmer organizations to create awareness about the policy makers and planners who have failed to have a vision on Indian agriculture in stead of providing cold storages and godowns these policy makers were busy in seeing to it that MNCs got profit. BKS State President Lingaraj Patil was present.

Source : Deccan Herald; May, 5, 2012,

56

Farmers Need to Unite

Belgaum - **May 5** - Bharatiya Krishak Samaj president Krishan Bir Chaudhary has urged farmers to get organised and fight for their rights, or they would continue to be exploited. Dr. Chaudhary, who participated in the farmers' conference in Bagalkot and was on his way back to New Delhi, was speaking to presspersons here. He said that despite protests at various fora the Union government was continuing its "anti-farmer" policies, resulting in 2.56 lakh farmers committing suicide in distress. "Had this happened in any developed country, several governments would have been dethroned," he said.

Opposing the opening of domestic markets to foreign players, he said allowing FDI in multi-brand retail would have disastrous consequences in the near future. Responding to questions, he said the government should work to uplift the people of the country, particularly farmers. However, he said the government would never change its mindset and therefore the farmers needed to organise to protect their rights. Dr. Chaudhary said the government should work on recharging the groundwater table and ensure optimum use of water.

Source : The Hindu; May 5, 2012;[http://www.thehindu.com/todays-paper/tp-national/tp-karnataka/farmers-need-to-unite/article3386158.ece]

57

Protect Farmers to Ensure Food Security

New Delhi - April, 22 - The government has failed to make agriculture a profitable enterprise and to raise the livelihood of the farmers, according to farmer leader and president of Bharatiya Krishak Samaj, Dr Krishan Bir Chaudhary. Chairing the national conference on "Food Security & Right to Food", organised jointly by FoodFirst Information & Action Network-India and Bharatiya Krishak Samaj, here, Dr Chaudhary said: "The government's policies are totally anti-farmer in nature and has failed to increase the income of the farmers. The continuing incidence of suicides by farmers bears testimony to this fact."

Sh. Harikesh Bahadur (Ex. M.P.), Dr. Krishan Bir Chaudhary (President BKS) Smt. Maya Singh (M.P.), Sh. Atul Anjan (Sect. CPI) on 21st April 2012.

Dr Chaudhary cautioned that unless and until the livelihood of the farmers improves and agriculture becomes a profitable enterprise, the food security of the country will be at peril. He particularly focussed on the issues of low agricultural productivity, environmental and health hazards of the genetically-modified seeds. He criticised the direct involvement of corporate houses in agriculture and said that the corporate houses are exploiting the farmers and paying them low returns. Dr Chaudhary said that the government should give top priority to soil health management and extension of irrigation facilities for increasing crop productivity. Incentives should be given for animal breed improvement programme for increasing the milk production. Mrs Maya Singh, MP from BJP, delivering the key note address welcomed the National Food Security Bill and said that the move is necessary in the face of prevalent hunger and malnutrition in India, but cautioned that the Bill might be a populist move in the absence of a holistic planning. The doling out of entitlements to a large number of population may boomerang and make the poor citizen a beggar, instead of a food secure person with dignity, she said.

Mr Atul Anjan stressed the need to implement the suggestions of the MS Swaminathan committee report on agriculture and to ensure judicious distribution of wealth. He also suggested that the MPLADS funds be abolished to stop widespread corruption in the scheme. Mr Harikesh Bahadur, former MP, said that the success of any government programme depends on the efficiency and good intention of the implementers. The moral and ethical values are now lacking in the society that is creating problems in every sphere of life and are also the reason for the failure of the government schemes.

Mr Ramesh Bhatt, Anchor (Lok Sabha TV) said that there are some serious challenges in the implementation of the National Food Security Bill like finalising the BPL list, ensuring viability of agriculture and availability of food grains, efficient public distribution system and sufficient budget for the implementation of the programme. Mr D Gurusamy stressed that the food security Bill promises several entitlements for the poor, but it should be treated as their right and not as a freebie. In Tamil Nadu, the free rice is black marketed for liquor making in Kerala, instead of being used for ensuring food security.

Source : The Statesman; April 22, 2012;[http://www.thestatesman.net /index.php?option=com_content&view=article&show=archive&id=407545&catid =42&year=2012&month=4&day=23&Itemid=66]

58

Convention report on Global Climate Change

New Delhi - July 28 - The Convention "Impact of Global Climate Change on Agriculture in Russia and India", organized by Bharatiya Krishak Samaj jointly with Russian Centre of Science & Culture in the capital was addressed by prominent academics, agricultural scientists, environmentalists, M.P.s, and others, and attended by a large gathering of agriculturists representing several Indian states. The speakers were unanimous in making a clarion call for joint efforts by Russia and India on reducing the negative impact of global climate change. On this occasion **Smt. Sonia Gandhi** Chairperson, UPA and **Dr. M. Veerappa Moily**, Union Minister of Law & Justice send their good wishes for the success of the programme.

In introductory address President, Bharatiya Krishak Samaj, **Dr. Krishan Bir Chaudhary** said that Climate change is a reality and the main cause of the present situation is on account of the anthropogenic activities disturbing the composition of the atmosphere resulting in higher concentration of Carbon Dioxide (CO_2) which accumulates along with other green house gases (GHG) like methane and nitrous oxide and contribute to increase in surface temperature of the earth. The main contributors have been the developed countries like US and EU but now other developing countries like China are slowly replacing as the main polluters. However, the per capita emission reveals that the main emitters are the developed countries. The consequences of these emissions are already visible with disturbance in climate which in turn is touching everyone's life. Climate is an important factor of agricultural productivity. Climate change is likely to impact agriculture and food security across the globe. In Another serious challenge confronting the agriculture is the competition for water resources increases, and the frequency of extreme temperatures changes.

From left – Maj. Gen. (Retd.) R.M. Kharab, Sh. Sunil Shastri, Sh. D.D. Lapang, Dr. Krishan Bir Chaudhary, Sh. Oscar Fernandes, H.E. Andrei A. Sorokin, H.E. Lt. Gen. M.M. Lakhera, Sh. Harish Rawat, Sh. Madan Lal Sharma, Sh. T. Meinya, Sh. Harikesh Bahadur

Voicing his concern on the negative impact of global climate change, the Chief Guest **H.E. Lt. Gen. (Retd.) M.M. Lakhera**, Governor of Mizoram, noted that the developing countries like India are highly vulnerable to its potential impact, adding that ironically the high-emission polluters in developed counties are going to be the beneficiaries of climate change and not its victims as far as food production is concerned. The world community needs to come together to discuss mitigation and adaptation strategies to counter global warming and climate change so that the poor are not made to carry the full burden of this man-made disaster, the Chief Guest said and added that what we need to do is to improve our traditional seeds in the Indian environment to achieve higher production by better means of water harvesting, soil fertility and organic fertilizers.

Earlier, welcoming the gathering, **Mr. Sergey Isaev**, Head of Science and Technology, RCSC, said that global warming is the observed increase in the average temperature of the Earth's atmosphere and oceans in recent decades and its project continuation into the future. He pointed out that Russia is today the world leader in reducing green house gas emissions. Russia accounts for half of all the reduction in emissions in the world over the last 20 years.

Making a clarion call on joint efforts by India and Russia towards reducing the impact of global climate change largely affecting mainly agricultural production, **Mr. Oscar Fernandes**, M.P., Chairman, Parliamentary Standing Committee on Human Resource Development, laid emphasis on focusing more on organic manure in agriculture, water conservation and water management. Globally, all societies will be vulnerable to changes in food production, quality and supply under climate change along with their consequent socio-economic pressures. Climate change is also expected to affect agricultural and livestock production, hydrologic balances, input supplies and other components of agricultural systems.

Pronouncing a note of warning that developed countries are more responsible for climate change, **Mr. Harish Rawat**, Minister of State for Labour and Employment, Government of India, said that those responsible should do the needful in the matter. In India, over 90 per cent of the people perform green job and they do not harm environment. Our Government is more concerned about agri-measures, and this is the major one among the eight missions it has launched, he noted.

Assessing the substantial climate change of recent years influencing all aspects of human life and activities, **H.E. Mr. Andrei A. Sorokin**, Charge d' Affaires, Embassy of the Russian Federation in India, pointed towards the green house gases and aerosol upsetting the radioactive balance of the system contributing to global warming. Citing the fact that Russia is one of the countries where agriculture depends largely on climate fluctuation, Mr. Sorokin said that the impact of climate change on agriculture in Russia is very complicated and little-investigated.

He expressed the hope that joint efforts of scientists from India and Russia will facilitate the introduction of innovative technologies and more effective cooperation in such critical areas as quality improvement systems for maintaining soil fertility and preventing land degradation, saving and mobilization of the gene pool of plants resources, effective biotechnologies for the selection of species with higher productivity and resistance to unfavourable environmental factors, the establishment of national systems of agro-ecological monitoring. Such cooperation could not only strengthen food security of our countries, but also make it possible to reduce the negative impact of global climate change, Mr. Sorokin concluded.

Maj. Gen. (Retd.) R. M. Kharab, Chairman, Animal Welfare Board of India, explained the negative effects creeping in the environment on account of human negligence and underlined various measures to obtain food security and better agricultural production.

Mr. Aboni Roy, M. P., in his observation referred to the imperative of stalling the efforts of developed countries in damaging the global climate with a view to maintaining the ecological balance.

Mr. Madan Lal Sharma, M.P., expressed the view that we in India are blessed with the greenery whereas the Western countries in abject violation of norms spoil the environmental harmony and balance.

Mr. Sajjan Singh Verma, M. P., called for more and more efforts to make our nature eco-friendly aimed it ensuring a sumptuous and comfortable climatic environment.

Mr. D. D. Lapang, Former Chief Minister of Meghalaya, Chairman, Nearth-East Congress Committee, said that climate change is a reality affecting agriculture and food security across the globe. He made an emphatic note on rising to the occasion to meet this challenging phenomenon.

Mr. Sunil Shastri, Ex M.P., described agriculture as the soul of the country's economy and agriculturists the backbone of our people. Referring to the global warming as a dangerous signal, he warned that consistent efforts are the need of the hour to contain it.

Mr. P. K. Thungan, Former Chief Minister of Arunachal Pradesh, underscored the great role India and Russia can play together joining other nations in confronting the serious issue of global warming thereby ensuring a harmless and safe atmosphere.

Mr. Harikesh Bahadur Ex. M.P., expressed the views that change of climate is effecting our agriculture productivity and the ground water level is going down.

Mr. Atul Kumar Anjan, Secretary, CPI, in his address was critical on the role of developed Western countries whose negligent attitude towards the crucial condition on global warming, which needs to be met with.

Dr. T. Meinya, M.P., in his presidential remarks pin pointed the suggestions and observations made by the distinguished speakers contributing to tackling of global warming so as to ensure ecological balance and a pure and peaceful environment.

Sh. Sangh Priya Gautam, former Union Minister , Smt. Neeva Konwar, Member, National Commission for Woman(Govt. of India), Sh. Oris Syiem Myriaw, Member National Commission for Scheduled Trides (Govt. of India), Capt. Praveen Davar & Sh. Ranji Thomas, Secretaries, All India Congress Committee, Jathedar Rachhpal Singh, Sh. K.Sareen, Sh. T.N.Fotedar, Sh. Sunil Dang, Sh. Ajay Gupta (ICCR), Sh.Hatam Singh Nagar, Gen.sect.(UPCC), Ch. Raghunath Singh, Prof. Kishore Gandhi, Dr. Sanjay Kaushik, Sh. Dhirendra Pratap-Uttrakhand, Sh. Manish Nagpal (Ex Minister, Uttrakhand), Ch. Mahabir Gulia , President, Haryana Krishak Samaj, Ch. Ram Karan Solanki President, Delhi Krishak Samaj, Ch. Bijendar Dalal, President, Palwal Krishak Samaj, Sh. Raja Matin Noori, Sh. Prahlad Tyagi, Sh. Manish Chaudhary (Debas), and other prominent persons attended the programme.

The following resolutions were passed in the convention

Key interventions needed to scale Indian agricultural challenges from climate change.

India has to takes on globally the climate change issues it needs to drastically reform its internal agricultural policy preparing itself in a war footing on mitigation and adaption. As part of the policy suggestion it was found that the following intervention would be needed immediately to equip Indian Agriculture to take on climate change:

1. Zero Tolerance to conversion of agricultural land for non-agricultural use.

2. A resolve to make few regions in India chemical and synthetic fertiliser free by 2020.

3. An urgent initiative or a bill to conserve biomass in the farm and Waste Recycling for Agriculture.

4. Incorporate in situ tree planting in all farming, adopt a Mixed farming as means to combat climate change.

5. Special Mission initiated at the Country level to shift crop acreage to Course Cereal and Millets to enhance nutrition value of food basket and help agriculture to adapt to climate change.

6. Free all the water bodies like ponds, lakes and tanks from illegal possession as per revenue record of every village and reforms initiated at the state level to rectify the same and scale up the level of water harvesting at a decentralized level.

7. Special Intervention from Indian Government to regulate the flood water for effective recharge using deep bore technologies at suitable depths.

8. Scaling up the organic agriculture and developing model centre of excellence and shift agriculture subsidies for intensive organic practices.

9. Revitalize the rural credit and crop insurance in the context of Climate change.

10. Launching of Sustainable Traditional Agricultural Revolution (STAR) using local resources for beating climate change.

Source : Kisan Ki Awaaz; July, 2010;[http://www.kisankiawaaz.org/Impact%20of%20Global%20Climate%20Change%20on%20Agriculture%20in%20Russia%20and%20India.html]

59

NBRA Bill will Silence Voices of Public Resentment

Mangalore - March 22 - The Union Ministry of Science and Technology is gearing up to table the National Biotechnology Regulatory Authority (NBRA) Bill in the Parliament.

The Ministry, which is promoting genetically modified (GM) crops, is unmindful of its health and environmental hazards revealed by a series of studies done by scientists across the world," said Bharatiya Krishak Samaj President Dr. Krishan Bir Chaudhary. Addressing media persons Dr. Chaudary said that the Genetic Engineering Approval Committee (GEAC) had not done proper homework before giving its nod for the clearance of Bt. Brinjal.

"Several lacunae were pointed out by eminent molecular biologist Dr Pushpa M Bhargava, who was a special invitee as per the Supreme Court directive, at the GEAC meetings.

But the GEAC, which is dominated by persons having vested interests, ignored Bhargava's suggestions and pushed for the release of Bt. Brinjal," Chaudhary alleged.

Through the proposed NBRA Bill, the voices of the public resentment will be silenced as it makes criticism of modern biotechnology an offence. Such law would not be keeping with the traditions of a vibrant democracy, he said. "Hazardous GM food should not be imposed on unwilling people through draconian laws. In a democracy, people should have the right to choose what they should eat," he added.

Reduces immunity

It has been scientifically proved that the consumption of GM food may reduce immunity system in human body. Further, it also would affect the digestive, metabolic functions and may cause carcinoma, Chaudhary explained.

"Genetic engineering is not the way to find solutions to the problems in agriculture sector. It is unfortunate that Indian government is promoting Bt. Brinjal even though the nation do not face shortage of it," he said and demanded that the states should ensure protection of seed rights of the farmers.

Now the people have come to know that there are two classes of agriculture scientists in India. While one class work for seed multinationals, another class pursue science in public interest, he said.

The objective of the seed multinationals is very clear that they want to control the entire food chain, he added.

Source : Deccan Herald; March 22, 2010;[http://www.deccanherald.com/content /59523/content/219337/content/218765/F]

60

Agriculture Development Needed for Prosperity

New Delhi - Feb 23 - The Bharatiya Krishak Samaj felicitated the Union minister for law and justice, Dr. M. Veerappa Moily on February 23, 2010 for penning two volumes of his book Unleashing India. The Volume 1 of Unleashing India deals with the theme - A Roadmap for Agrarian Wealth Creation. The Volume II of Unleashing India deals with the theme - Water: Elixir of Life. Both these volumes are helpful for policymakers for drawing up policies for the agriculture sector.

Dr. Moily, who is a seasoned politician, former chief minister, former chairman of Administrative Reform Commission and experienced administrator speaking on the occasion at Speaker's Hall in the Constitution Club called for upliftment of farmers. He said that a civilization can thrive as long as farmers and agriculture exist. Development depends upon the development of agriculture sector. The great Indus Valley civilization vanished as agriculture sector suffered heavily.

He, therefore, called for increasing the livelihood of farmers, increasing agriculture production and conservation of water. Dr. Moily said : "I would like to sit and discuss with Bharatiya Krishak Samaj on important issues relating to farmers. I know Samaj is doing a lot for voicing the concerns of Indian farmers." He inaugurated the Website of Organisation vide www.kisankiawaaz.org

The President of Bharatiya Krishak Samaj, Dr. Krishan Bir Chaudhary brought to the notice of the minister the plight of the farmers in the backdrop of rising input costs and low returns for farm produces. He assured to cooperate with the UPA government in finding out solutions for the benefit of farmers. He appreciated the UPA government's sincerity in helping farmers.

The former Governor of Uttar Pradesh, Mr. Romesh Bhandari called for measures for increasing the livelihood of farmers. Mr. T. Mainiya (M.P.), Mr. Bapi Raju (M.P.), Mr. R.C. Khuntia (M.P.) Chaudhary Lal Singh (M . P .) & Parliamentarians Mr. Sunil Shastri, Mr. Aslam Sher Khan , Mr. Harikesh Bahadur, Mr. D.P. Yadav, Mr. P.K. Thungan, Mr. Virendar Kataria (Former President, Punjab Congress Committee), Sqn. Ldr. Mr. Kamal Chaudhary, Mr.A.Sangtan, Mr. R.R. Sahu, Mr. Vishnu Prasad, Mr.D. Kalidas & Secretaries, All India Congress Committee Capt Praveen Davar, Mr. Ranji Thomas & Senior Congress leaders, Sardar Rachh Pal Singh, Mr.T. Sering Samthel (Member, Tribal Commission, Govt. of India) Mr. Purushottam Goyal (Ex Speaker, Delhi), Mr. lalit jain(Ex. Minister MP), Mr.K.Sareen, Sardar Harcharan Singh Josh,Ch.Ram Karan Solanki (Palam) and several others leaders appreciated the efforts of the government. and said that the UPA government would render all possible help to farmers.

Dr. T. Mainya (M.P.), Dr. M. Veerappa Moily, Minister for Law & Justice, Govt. of India Dr. Krishan Bir Chaudhary, President (BKS), Ch. Lal Singh (M.P.)

Source : Kisan Ki Awaaz; March, 2010;[http://www.kisankiawaaz.org/ Agriculture%20development%20needed%20for%20prosperity%20-%20Dr.Moily.html]

61

Bill On GM Crops Opposed

Mangalore - **March 23** - Dr. Krishan Bir Chaudhary, president of the Bharatiya Krishak Samaj, has expressed concerns over the National Biotechnology Regulatory Authority Bill, which is likely to be tabled soon in Parliament.

Mr. Chaudhary, a critic of genetically modified crops, who was here to participate in a recently concluded seminar on biodiversity organised by the Nagarika Seva Trust, was addressing presspersons before his departure on Monday.

MAKING A POINT:Krishan Bir Chaudhary, president of the Bharatiya Krishak Samaj, New Delhi, at the press conference in Mangalore on Monday.

'It is against the principles of federalism'
'It seeks to silence critics of biotechnology regime'

The Bill, which seeks to set up a single national-level regulatory body with exclusive authority over the release and control of genetically modified crops in the country, was described by Mr. Chaudhary as "draconian". Claiming that the basic premise of the Bill was flawed, he said that it was against the principles of federalism that envisions agriculture as a State subject. (According to Section 81 of the Bill, the Act will have an overriding effect) Pointing to a clause in the Bill that penalises public criticism of GM foods and crops, he said the new Bill sought to muzzle the critics of the biotechnology regime.

The clause, which is titled "Misleading public about (genetically modified) organism and products", advocates the imposition of a penalty, including six to 12 months of imprisonment and (or) a fine of up to Rs. 2 lakh.

According to him, the most disturbing aspect of the Bill is that the Biotechnology Regulatory Authority will have sole powers to enforce the provisions of the Act and that it will operate outside the purview of the regular courts.

Calling for a national referendum on GM crops, he said that the Indian Council of Agricultural Research (ICAR) had become prone to influence from "extraneous and alien forces".

Source : The hindu ; March 23, 2010;[http://www.hindu.com/2010/03/23 /stories/2010032362030300.htm]

62

MNCs will Dominate if Seeds Bill Adopted

New Delhi - June 29 - The farmers' organisation Bharatiya Krishak Samaj has raised concerns over the Seeds Bill saying it would increase the domination of multi-national seed companies in India and may force farmers to pay royalty on hybrid seeds.

"The Indian farmers will lose their rights on using seeds of their choice and it would mainly promote interests of the multi-national firms," BKS President Dr. Krishan Bir Chaudhary said.

The bill would serve the interests of firms producing genetically modified (GM) seeds in the country, he added, saying the bill may force farmers to pay royalty on hybrid seeds.

The government tabled the controversial Seeds Bill in the Rajya Sabha in December 2004 and later it was referred to the parliamentary standing committee on agriculture for review.

The standing committee took about two years to review the Bill and had submitted its report in 2006. The report is still pending with the government and is expected to be intoduced in the coming session of Parliament.

Chaudhary said the bill would also jeopardise the country's food security. "By continous use of hybrid seeds, the farmers would be gradually obliged to buy seeds from the MNCs," he warned. There is no traditional seed for cotton available in the market, he added.

Chaudhary said that the real motive of the bill is not to provide quality seeds to the farmers. Instead it could result in scarcity of natural seeds, he said, adding that it would lead to a lot of litigation as multi-national seed firms can claim intellectual property rights to the seeds that farmers use.

On the litigation issue, Chaudhary added that as per the World Trade Organisation obligations, India had passed Plant Variety Protection and Farmers Rights Act (PVPFRA) under which interests of breeder and farmers have been protected.

He said the bill, if passed by the Parliament, would nullify the traditional rights of farmers given by PVPFRA on seeds. "It will also increase the domination of multi-national seed companies on the Indian seed market,"

He added the government has by and large sidelined the recommendations of parliamentary standing committee.

Chaudhary said in India 'Bt. gene' is being used in hybrid varieties, but in countries like China it has only been used to improve the natural seeds so that farmers are not dependent on private companies for hybrid seeds always.

Source : Press Trust of India & Business Standard; June 30, 2009;[http://www. business-standard.com/article/economy-policy/mncs-will-dominate-if-seeds-bill-adopted-farmers-associations-109062900135_1.html]

63

Spurious Pesticide Kills Crops worth Rs 25,000 crores every year

New Delhi - April, 6 - India is facing crop losses worth over Rs 25,000 crore a year because of the use of spurious pesticides and insecticides, a leading farmers' body claimed here on Monday, pegging the annual sale of such fake products at over Rs 1,500 crore.

Calling for a crackdown on the companies selling spurious items, Bharatiya Krishak Samaj president Krishan Bir Chaudhary told reporters that out of four widely used products in northern India, manufactured by at least 14 companies, as many as 11 companies failed to meet the parameters set by the government.

"Crops worth Rs 25,000 crore are being lost because of spurious pesticides and insecticides that notch up annual sales of about Rs 1,500 crore," he said. "About 30% of the sugarcane crops in the second-largest sugar producing state of Uttar Pradesh are lost due to such fake products."

The farmers' body carried out a random sampling of pesticides and insecticides in different places and sent those for tests to the Institute of Pesticide Formulation Technology, which has been accredited by the National Accreditation Board for Testing and Calibration Laboratories under the ministry of science and technology.

In the manufacturing of phorate, 10% CG which is widely used in sugarcane, paddy, wheat and potato crops, samples of 10 out of 11 companies failed to meet the required standard, according to Mr. Chaudhary. Similarly, the samples of five companies of isoproturon, 75% WP, which is used against insects to protect paddy and wheat crops are alleged to have failed in the test.

Samples of two out of four companies tested are claimed to have failed in complying with requirements for the manufacturing of 2,4-d ethyl ester, 38% EC, which is used in the paddy crop.

Moreover, three companies are alleged to have bypassed norms as reflected in the samples of 2,4-d sodium salt, 80% TECH, which is used in the paddy crop, he said.

Mr. Chaudhary said the government should clamp down on products of companies from time to time and cancel licences of erring firms while taking action against pesticide producers and inspectors who connive with the manufacturers.

Sources: PTI, The Hindu, The Economic Times, The Financial Express; April, 6, 2009

64

Farming Matters

New Delhi, March, 17 - Producers attended a large meeting of the national farmers' organizations this February, condemned the corporate model of agriculture being promoted across India, asked for urgent adoption of farmer-centric policy by the Government and the withdrawal of GM crops on health grounds.

Addressing the gathering, the organization's president Dr. Krishan Bir Chaudhary highlighted that current Indian agricultural policies are causing serious suffering for many farmers, and even leading to a high rate of suicide among those working on the land.

Dr. Chaudhary also called for action to mobilize youth as an immediate strategy to beat recession —with a focus on providing technical and financial assistance for agro-based small-scale productions to assist unemployed rural youths.

The farmers' leader was also critical about the existing policy of the Special Economic Zones (SEZs) and called for these areas to only be set up on barren land, and not on fertile regions which can supply food.

Chaudhary also expressed concern over the rapid spread of colonialization and India's loss of food security in the hands of multinational corporations.

'More than 1.6 lakh (hundred thousand) farmers committed suicide after India signed the WTO agreement', he said.

'When the country already has Seed Act 1968, what is the relevance of introducing one more Seed Bill in the Parliament? It is nothing but auctioning our original rights to the multinationals.'

Source : www.deccanherald.com, www.slowfood.com; March 17, 2009;[http://www.kisankiawaaz.org/Farming%20Matters.html]

65

Farmers happy over Doha Failure; want WTO's wind-up

New Delhi - July, 31 (UNI) - Indian farmers are relieved over the collapse of the global trade talks at the World Trade Organisation. Stating that the failure of the 9-day-long mini-ministerial held in Geneva to finalise the Doha Round has vindicated the stand of Bharatiya Krishak Samaj, its president Krishan Bir Chaudhary said WTO has become "ineffective and needs to be wind up".

He said; "WTO is practically hopeless and helpless after successive failures of a series of attempts to revive the multilateral trade negotiations. The time has already demonstrated that this multi-lateral trade system is absolutely incapable of achieving the objectives of the Doha Development Round.

Therefore, the Bharatiya Krishak Samaj has urged that WTO should be wind up immediately".

"We are very happy. Indian farmers have always wanted agriculture to be taken out of the Doha Round talks,"

Chaudhary said the latest WTO text on agriculture completely ignored the food security and livelihood concerns of the farmers in the developing countries.

The draft, he added, proposed a weak defence against the

possible influx of cheap and subsidised imports by suggesting a complicated system for implementation of special safeguard mechanism (SSM) by the developing countries.

The BKS complimented Mr Kamal Nath for taking a firm stand on the SSM.

He said the collapse is also good for the infant industries as the revised Nama (industrial products)draft proposals, if accepted, will have also sounded death knell for the small and medium industries in the developing countries.

Source :   UNI & One India News; July 31, 2008; [http://news.oneindia.in/2008/07/31/farmers-happy-over-doha-failurewant-wtos-wind-up-1217511328.html]

66

The Right to Protest

New Delhi - March 30 - The European Patent Office (EPO) has opened a Pandora's box by deciding to grant patent rights to seeds developed through conventional breeding processes. They have begun to grant both product and process patents for these, and as an interim ruling, EPO's Enlarged Board of Appeal (EBA) has decided to grant a general patent on broccoli.

The development is of special concern to India because farm exports from the country, as also from other developing countries, to Europe would be at stake, as such a measure may lead to patenting of a large number of crops. Trade disputes may, therefore, become inevitable, according to experts.

The immediate reaction to the EPO's plan to grant general patent rights to conventional crops has been the collective protest by farmers from India, Europe and Latin America, who have gathered in Munich, where the EPO is based. The Indian farmers are represented by the Bharatiya Krishak Samaj.

In a telephonic conversation from Munich, the President of B.K.S. Dr. Krishan Bir Chaudhary, said, "We will fight for our sovereign rights over seeds and farm animals. We cannot afford to lose our rights to MNCs.

We know the strategy of MNCs like Monsanto, which sued Canadian farmer Percy Schmieser for ownership of his conventional canola seeds, after having his field contaminated by pollen from the nearby genetically modified (GM) canola fields."

The EPO's decision to accord such patent rights flows from the ruling of the EBA, which is also to decide on the validity of a patent on broccoli (EP 1069819 B1) this year. Since 1980, the EPO has granted patent rights to over 151 GM crops from the 285 applications it received. It has now on its agenda the grant of patent rights to conventionally bred seeds and animals.

The farmers have been joined by NGOs like Greenpeace, Misereor, Swissaid, The Declaration of Berne and No Patents on Life, and have issued a global appeal against the EPO's decision. On April 28, the EPO rejected an application for a patent on sunflowers derived from normal breeding (EP1185161), which was filed by Greenpeace. "We will file another appeal against the EPO decision, as according to the European Patent Convention, conventional plants cannot be patented, only GM crops can be," said Christoph Then of Greenpeace.

While patent claims have been made for soybean, the most threatening example is of Syngenta, which has claimed patent rights over a large sequences of rice genomes and is also extending its rights over genomic information of other crops.

The EPO has already granted a patent right to a Dutch company, Rijk Zwaan, on aphid-resistant plant composites (lettuce), and is slated to decide on patent rights for a method of increasing a specific compound in Brassica species. Monsanto too has claimed patent rights over pig breeding process. With such broad patents likely to be granted, Indian farmers' groups feel it may jeopardize the country's trade interest as well as farmers' sovereignty over seeds.

Source : The Indian Express; March 30, 2007; [http://www.indianexpress.com/news/the-right-to-protest/26990/]

67

Make Changes in Seed Bill, Urge Farmers

New Delhi, Dec. 3 - Farmers have urged the government to follow the Parliamentary panel's suggestions and effect necessary changes in the Seeds Bill, 2004 for ensuring farmer rights. The government tabled the controversial Seeds Bill in the Rajya Sabha in December 2004, later it was referred to the parliamentary standing committee on agriculture for review. The standing panel took about two years to review the Bill and recently submitted its report.

Krishan Bir Chaudhary, head of the India's farmers' organization who deposed before the House panel is, however, apprehensive. He feels the government may bypass the panel's recommendations due to pressure from the industry. "It would be regrettable if the government ignores relevant suggestions made by the peoples' representatives and opts to favour the seed industry. If the government does so then we will launch a nationwide agitation to restore our rights," he said.

The House panel headed by Ram Gopal Yadav strongly recommended that the farmer's right to exchange unbranded seeds among themselves be acknowledged. It said, "the Plant Varieties Protection & Farmers' Rights (PVP&FR) Act 2001 should be made fully operative first," before implementing the proposed Bill. The new law should not undermine the provisions of PVP & FR Act.

In this context, Chaudhary said: "On behalf of the farmers I raised the issue of deleting the latter portion of Clause 43 (1) of the Bill so as to exempt farmers from any norms for saving and exchanging unbranded seeds.

The panel agreed to my suggestion that the latter portion of the clause is restrictive and should be deleted."He further said the PVP&FR Act should be implemented immediately and added, "In my personal view farmers should have the right to save and exchange whatever seeds they cultivate."

Chaudhary said the House panel also said the Bill should address the issue of promoting sale of newly developed seeds, and at the same time, ensure that farmers' interests and sustainability of agriculture are not jeopardised. The law should strengthen the integrated growth of farmers and seed systems so that every farmer has access to quality planting materials at reasonable prices.

He said that his behest the panel also considered the issue of genetically modified (GM) seeds and suggested that the new law should not undermine the existing procedure for assessment and release of GM seeds. It criticised the proposed provisional registration of GM seeds and also self-certification of seeds by seed companies as this would facilitate backdoor entry of unapproved GM seeds.

On pricing of seeds, the panel report said that the proposed Central Seed Committee should be given adequate powers to regulate seed prices. It cited "unfair trade practices" committed by the seed multinational Monsanto in charging exorbitant prices for Bt cotton seeds.

The panel suggested inclusion of a clause in the Bill for compensating farmers for losses due to bad quality seeds on the line of that in the PVP&FR Act and a designated arbitration tribunal or compensation committee be constituted for the purpose.

It called for a Seed Mark to denote quality of seeds and labeling norms in harmony with the Weights & Measures Act and Packaging Act Seed inspectors should not search farmers.

Among others who deposed before the House panel were

Vandana Shiva of Navdanya, Suman Sahai of Gene Campaign, S Ramachandran Pillai and K Varadharajan of CPI(M)' farmer's wing, Kolli Nageswara Rao and Chittar Singh of CPI's farmers' wing, Manavendra Kachole, Nikhade and Govind Joshi of the Pune-based Shetkari Sangatana and representatives of the seed industry.

Source : The Financial Express; December 04, 2006; [http://www.financialexpress.com/news/make-changes-in-seed-bill-urge-farmers/185549]

68

Farmers' wing against Wheat Import

New Delhi- May, 9 - Taking the cue from Congress president Sonia Gandhi cautioning Prime Minister Manmohan Singh not to rush headlong with signing of free trade agreements (FTAs), the farmers organisation lashed out at unilateral liberalisation policy of government aimed at greater involvement of corporate houses.

The leader of the country's largest farmers' organisation, BKS, Dr. Krishan Bir Chaudhary criticised the government, particularly agriculture minister Sharad Pawar for going ahead with import of 3.5 million tonnes of wheat when the granary is full. He alleged that it is a deliberate attempt to erode country's self-sufficiency in food.

He said that the government policies of the day are no longer based on ensuring food security and farmers livelihood, but are dictated by WTO, World Bank, IMF and USDA, and are best suited to serve the interests of corporate houses and multinationals.

Mr. Chaudhary said, " It is shame on the part of the government to take the pretext of rising domestic prices of wheat to make a case for imports : There is enough stock in the country, with a wheat production of 72 million tonne in 2005, and expected 73.1 million tonne, this year. In the current season, the area under wheat has increased by 4 lakh hectare."

He said that the uptrend in domestic wheat prices is due to large scale hoarding by traders and multinationals. This is due to the removal of restrictions on stocking.

The government should immediately re-impose stocking limits to check market manipulations, he said.

Mr. Chaudhary criticised the involvement of corporate houses in direct wheat marketing, and said that they are paying more to the farmers taking advantage of the low minimum support price fixed by the government.

In the long run the corporate are not going to pay farmers lucrative prices and would buy the farmers' produces at distress sales, as has been the case with African cocoa growers, he said.

Source : The Financial Express; May,9, 2006; [http://www.financialexpress.com/news /farmer-groups-lash-out-at-govt-policies/167215/0]

69

Farmer bodies of UPA against the Government

New Delhi- Feb. 20 -The farmer organisations have cautioned the Congress-led government against its "anti-farmer policies". They have warned that unless the ruling UPA coalition mends its ways it may face the same fate as that of the erstwhile NDA government at the Centre.

Incidentally, such voices of resentment have come from the farmers' outfit of the ruling UPA partners. Krishak Samaj (BKS), the farmers' outfit of the Congress party, has called for immediate withdrawal of the amendments to the Seeds Act tabled in Parliament.

Speaking to the FE Krishak Samaj President Dr. Krishan Bir Chaudhary said: "The amendments proposed to the Seeds Act are anti-farmer as they calls for mandatory registration of seeds and prevents farmers from saving seeds for the next crop.

The Plant Varieties Protection & Farmers' Rights (PVP&FR) Act is sufficient to regulate the seed sector. There is no need for any other legislation. Rather the PVP&FR Act should be further amended in the interests of farmers and for making biopiracy impossible."

Dr. Chaudhary criticised the prime minister and the agriculture minister for suggesting that farmers diversify out of wheat and rice cultivation when the government, apprehending shortage, has planned to import five lakh tonnes of wheat.

He said there is no need to import wheat as there is enough stock in the country. He alleged that artificial shortage has been created due to hoarding by traders and multinationals like Cargill and ITC.

Dr. Chaudhary also criticised the proposed Indo-US intitiative in Agriculture Research and Education, describing it as against the interests of Indian farmers and a waste of public funds.

He said that transgenic technology worldwide in crops has created health and environmental hazards and consumers are reluctant to consume GM foods.

The US, through this collaboration, intends to thrust GM food on Indian consumers. The US and multinationals will gain free access to the genetic biodiversity and indulge in biopiracy, he said.

The general secretary of CPI's All India Kisan Sabha, Atul Kumar Anjaan, in addition, said, "As per the proposed agreement Indian scientists who would be pursuing studies in US under the exchange programme will have to pay a hefty fee of Rs 400 crore."

He criticised the government's move to invite FDI in retail sector as damaging to the interests of farmers and small retailers. The joint secretary of CPM's farmers' oganization, NK Shukla said: "By this agreement, US would extend a strong patent regime in agriculture, affecting the interests of Indian farmers."

Source : The Financial Express; February, 20, 2006; [http://www.financialexpress.com/news/story/147377]

70

Farm wing hits out at WTO Ministerial

New Delhi- Dec. 26 - The farmer's outfit of the ruling Congress party missed no chance in criticising the government over the deal it has struck at the recent WTO ministerial at Hong Kong.

The President of BKS Dr. Krishan Bir Chaudhary, who recently returned from Hong Kong said, "Before the ministerial I had written to our Congress president and chairperson of the National Advisory Committee, Sonia Gandhi to advise the government not to compromise on the interests of farmers. But matters took a different turn.

India and the other developing countries after compromising on Nama and services sectors could not gain in agriculture."

Mr. Chaudhary blasted the Union commerce minister, Kamal Nath for making tall claims over "gains in agriculture." He said that the Hong Kong declaration has caused more harm to the agriculture in the Third World countries.

The European Union should have phased out export subsidies long back. The Hong Kong declaration rather legitimised European export subsidies till 2013.

"As the European export subsidy would continue till 2013, the US would not be inclined to phase out its support to its state trading enterprises and would continue dumping food as aid.

The US had already said that it would phase out its support to state trading enterprises and stop dumping food as aid, provided Europe ends its export subsidy," said Mr. Chaudhary.

He said that the Hong Kong declaration also legitimised US cotton subsidy till 2006. The US cotton subsidy was slated to be phased out by September, 2005 as per the ruling of the WTO dispute settlement body.

He added: "The European export subsidy is a small part of the total subsidy given by the rich. As the ministerial has failed to phase out export subsidy, hopes are dim for future negotiations to arrive at a drastic cut in the heavy subsidy bill of the developed world."

Mr. Chaudhary cautioned that the para 24 of the ministerial declaration has tied up agriculture and NAMA together for negotiations on market access.

This, he said, is a dangerous move as it would in future give scope for more trade off deals, jeopardising the interests of farmers, labourers and industry in the Third World.

Source : The Financial Express; December, 26, 2005;[http://www.financialexpress.com/ news/farm-wing-of-congress-hits-out-at-wto-ministerial/154234/1]

71

Farm Union Warns Against
Sell out at WTO

New Delhi- Dec. 10 - The farmers' outfit of the ruling Congress has cautioned the government against striking any unfair deal at the upcoming WTO ministerial in Hong Kong. Rather, it says, India should negotiate to rectify the wrongs done to farmers in the Third World because of unfair trade practices.

The President of BKS Dr. Krishan Bir Chaudhary, in a letter to Congress president and chairperson of the National Advisory Council, Sonia Gandhi said, "The government would face dire consequences if it goes ahead to sign a deal which would be detrimental to the interests of Indian farmers."

Refering to commerce minister Kamal Nath's recent assurance that India would not allow sacrificing of farmers' interests in exchange for a better deal in sectors like NAMA and services, Dr Chaudhary said: "I hope the minister would live up to his assurances."

Dr. Chaudhary, who is on his way to Hong Kong to represent the country's farmers, drew Ms Gandhi's attention to the fact that for the past 10 years the developed countries have not implemented their commitments to reduce their high levels of farm subsidies and support.

Rather they have increased their levels of support and managed to shift their subsidies from one box to another. These practices have depressed global prices and resulted in cheap dumping of agro produces in the Third World, he said.

In such a situation, Dr. Chaudhary suggested that India should take the leadership in asking for restoration of the mechanism of imposing quantitative restrictions (QRs) on imports by the developing countries until the developed countries drastically reduce subsidies from the "applied levels" and make deep cuts in their tariff barriers.

He said that India and other developing countries should not make any commitments to reduce their farm tariffs as this is the only available weapon to check the floods of cheap subsidised imports.

He said if at all there is an issue for developing countries to reduce their farm tariff, this should be done from the bound tariff levels and not from the applied tariff levels.

For developed countries it should be from the applied tariff levels. He said that all subsidies presently rendered by the developed countries are trade distoring.

These subsidies should be brought under one Box with a view to prevent shifting of subsidies from one box to another.

Source : The Financial Express; December, 10, 2005; [http://www.financialexpress.com news/farm-union-warns-against-sellout-at-wto/156521]

72

Farmers Hail PVP & FR Act Notification

New Delhi- Nov. 14 - Farmers have hailed the government's decision to notify the Plant Varieties Protection and Farmers' Rights (PVP&FR) Act, 2001. They said that though the decision is belated, it would solve the farmers' problems to a great extent.

The legislation was passed by Parliament way back in 2001 and received Presidential assent in the same year, but was withheld from notification, which prevented its implementation over the past few years.

The Act, apart from protecting farm bio-diversity, allows farmers to save and exchange seeds in unbranded form for use in the next crop season. The Act has also banned registration of seeds containing terminator technology vide section 18 (1) (C).

The government has recently constituted Plant Varieties Protection and Farmers' Rights Board under the chairmanship of Dr S Nagarajan for implementation of the Act.

Speaking to FE, President of Krishak Samaj Dr. Krishan Bir Chaudhary said:

The PVP&FR Act was long withheld from its implementation due to pressure from the interested lobby of seed companies. This law gives some leverage to farmers in matters of use of seeds, though not complete freedom. There had been recent attempts to nullify this meagre freedom given to farmers under this Act by the introduction of amendments to the Seeds Act in the Parliament.

The farmers will not tolerate any such move and demand immediate withdrawal of the proposed amendments to the Seeds Act.

Rather PVP & FR Act should be further amended to give more freedom to farmers.

Dr. Chaudhary had earlier expressed concern over Delta & Pine Land announcing its new plans to foray into the seed sector, after it got patent rights for its terminator technology in patent offices abroad.

He now said that with the notification of PVP &FR Act, the country is better poised to deal with this situation.

Dr. Chaudhary was the sole farmers' representative to the recently held global conference on biotechnology hosted by Asia-Pacific Association of Agriculture Research Institutions (APAARI) and FAO in Bangkok where he demanded that the seed multinationals compensate farmers for the failure of Bt cotton.

Source : The Financial Express & www.nwrage.org ; November ,14 2005;[http://www.financialexpress.com /news/farmers-hail-pvp-amp-fr-act-notification-/159181/0]

73

Farmers Concerned over Delta & Pine Land Terminator Patent

New Delhi- Oct. 28 - Indian farmers have expressed grave concern over the patent rights accorded to Delta & Pine Land in Europe and US over its controversial terminator technology.

They have expressed fears that the company which has recently declared that it would foray into the country's farm sector in big way, may bring in the terminator technology. This terminator technology is detrimental to the interests of farmers, they said.

Indian farmer leader, Dr Krishan Bir Chaudhary said: that the government should take immediate steps to ban terminator technology in the country. It should immediately review the activities and intentions of Delta & Pine Land.

The company should not be allowed any field trials of terminator seeds. The pollen flow from plants with terminator technology to other crops will have dangerous consequences. It would make the pollen-affected crops sterile.

Mr Chaudhary said that it the hidden agenda of the corporate houses is to monopolise the seed sector. It is for this reason the seed companies are producing hybrid seeds which the farmers cannot save for the next season.

They usually do not produce conventional varietal seeds which the farmers can save for the next season.

Now with the terminator technology, the seed companies intends to complete their agenda of monopolising the entire seed sector as the plants of terminator technology will produce only sterile seeds, Mr.Chaudhary said.

Greenpeace has recently exposed the details of the patent for the controversial terminator technology granted in Europe on 5 October 2005.

The terminator patent has been approved for all plants that are genetically engineered so that their seeds will not germinate.

Source : www.nwrage.org & The Financial Express ; October, 28 2005;[http://www.nwrage.org/content /farmers-concerned-over-dpls-terminator-patent]

74

India's Seed Bill may be Delayed

New Delhi- April, 11 - The Union agriculture ministry's attempt to rush the draft National Seed Bill may be delayed. Acting on a representation made by an apex farmers' organisation, the chairperson of the National Advisory Council, Sonia Gandhi, directed the ministry to reconsider certain clauses in the proposed draft that are likely to hamper the interests of farmers.

Incidentally, such an opposition comes from the farmers' groups of the ruling Congress party, Bharatiya Krishak Samaj (BKS). Though the BKS President, Dr. Krishan Bir Chaudhary had made a representation to Ms Gandhi who is the chairperson of the UPA coalition and president of the Congress party, she chose to act on this issue as the chairperson of the National Advisory Council to the government and not as a key force behind the government.

The BKS leader, in his letter to Ms Gandhi, had said: "The bill is a clear trap to curb the traditional and indigenous rights of our peasantry to grow, breed, multiply, preserve and exchange seeds.

The seed bill is wholly incongruous. Sinister as it is, it will demolish the time tested agrarian culture and the socio-economic fabric of the rural India that has for centuries worked faultlessly and sustained our small and marginal farmers, having even less than two acres of land.

83% farmers use their own farm-saved seeds.

In one stroke, the National Seed Bill on enactment will reduce 36 crore farming families into pathetic non-entity and make them captive at the mercy of seed multinationals, aided and abetted by the unabashed and insensitive state machinery.

The draft bill makes registration of seeds mandatory and in this context, Dr. Chaudhary in his letter said: "The National Seed Bill treats farmers as traders. They will be hounded to run about for registration if they grow and exchange seeds." Dr. Chaudhary not only opposed the draft bill but also the introduction of transgenic seeds.

He said, "Seed is the most vital factor in enhancing agricultural production. The National Seed Bill should not put any infringement on the indigenous and traditional rights of the farmers to grow, breed, multiply, exchange and store seeds and be prevented to carry on the age old and time-tested barter system for mutual benefits of the fellow farmers.

Farmers should not be treated as traders in the proposed bill. The government should bar the access of transgenic seeds and terminator technology in our agro- system for all times to come.

He also called for remunerative minimum support prices for crops and cautioned the government not to dismantle the state-sponsored procurement of grains, encouragement of organic farming and post-harvest management.

Source : The Financial Express; April,11, 2005; [http://www.financialexpress.com/news /draftofseedbillmaybedelayed/131141/0]

75

Illegal Patent of Indian Wheat Revoked

New Delhi- Oct. 10 - The organisations fiercely seeking revocation of illegal patenting of Indian agricultural products emerged victorious as Munich-based European Patent Office (EPO) acted upon their petition and revoked in "total" a patent on Indian Wheat variety recently.

"This victory will go a long way to save the rights of the farmers in India," they said. The multinational corporation Monsanto was unable to file a contest against the petition challenging the patent in its favour.

The three organisations Bharatiya Krishak Samaj (BKS) and Navdanya & Greenpeace had taken the initiative to fight for Indian farmers to save Indian wheat variety Naphal which was patented in EPO in Munich, Germany.

The fighters against this bio-piracy had unitedly filed their opposition challenging the unjustified patenting of the product in EPO on January 27, 2004. The patent on an Indian wheat variety was granted to the Monsanto on May 21, 2003.

BKS president Krishan Bir Chaudhary said Monsanto got this patent by transferring genes from an Indian wheat landrace Naphal to their Galahad-7 wheat variety. They claimed this act of theft as an "invention".

Dr Chaudhary said, "This patent was a blatant example of bio-piracy as it forfeited the rights of Indian farmers and dismissed hundreds of years of hard field work of Indian peasantry.

Source : The Tribune; October,10, 2004;[http://www.tribuneindia.com /2004/20041011/biz.htm#1]

76

Three NGO's File Petition Against
Wheat Patent for Monsanto

New Delhi- Jan. 28 - International non-government organisation Greenpeace along with the Indian farmers organisation (BKS) and Research Foundation for Science, Technology and Ecology (RFSTE) filed a petition at the European Patent Office (EPO), Munich challenging the patent rights given to Monsanto on Indian landrace of wheat, Nap Hal.

Patent expert Christoph Then of Germany & Aseesh Tayal of Greenpeace-India, Krishan Bir Chaudhary of BKS and Vandana Shiva of RFSTE jointly signed the petition.

They said that the patented variety of wheat with specific baking characteristics of flour derived from it was orginally developed in India and has been cultivated, bred and processed for bread (Chapatis) by Indian farmers for years.

The patent that Monsanto now holds means it has the monopoly on farming, breeding and processing of this type of wheat. The patent (EP 445929) was granted on May 21, 2003 by EPO, Munich.

According to the European Patents Convention, patents cannot be issued on plants that are normally cultivated, any more than they are allowed to be issued on their seeds.

In case of the Monsanto wheat patent, the EPO has clearly disregarded rules and law, the press release said.

Dr. Aseesh Tayal of Greenpeace-India said "the patent is a blatant example of biopiracy as it is tantamount to the theft of the results of the breeding efforts of Indian farmers.

Unless we are able to successfully subvert the attempts of the multinationals, Indian farmers and scientists could end up paying royalties to foreign corporations for using traditional knowledge that their own communities have developed."

Our investigations show that the examiners at the EPO knew that the wheat involved is cultivated entirely normally and that it is not an invention, Monsanto employed all kinds of tricks and deception to conceal this fact, though the truth is amply clear on proper examination.

This case shows again that only a clear prohibition of patents on plants and seeds can stop the ongoing abuse of modern patent law, said Dr. Christoph Then, a patent expert of Greenpeace German office.

Dr. Krishan Bir Chaudhary of Krishak Samaj called upon European governments, particularly the German government to prevent such recurrences in future.

Source : The Financial Express; January, 28, 2004;[http://www.financialexpress.com/news/3-ngos-file-petition-against-wheat-patent-for-monsanto/97965]

77

Cultivating Ideas

Garda Italy - July 15 - Indian farmer's leader is not in Italy for the Grand Tour. He is representing millions of farmers who are sinking into their native India's well of rural poverty and he wants to tell the rest of the world exactly what is going on in India. Dr. Krishan Bir Chaudhary is speaking on behalf of the Indian farmers. He is the President of Indian farmers' organization, an association representing India's small farms.

Giovanni Alemanno, Italy's agriculture minister & Dr. Krishan Bir Chaudhary in Slow Food's National Congress held in northern Italy on 6 July 2002

On his way to the UN's ill-fated World Food Summit in Rome (July 8-10, 2002), Chaudhary stopped at Slow Food's National Congress held in northern Italy, to talk about what the farmers of India want and what they don't want, which is, in a Chaudhary nutshell, 'control over their own land and no more genetically modified crops'. In an interview Chaudhary spoke of their fight for a return to biodiversity and Indian farming practices.

Chaudhary's agenda is to improve the conditions of India's farmers; he represents small and marginal farmers (those with less than two acres holding). Citing its failure to lower the price of seeds, implement regulations against GM introductions, lower prices for pesticides, diesel, electricity and farm machinery, he calls the government anti-farmer and is calling for change. The answer, Chaudhary believes, is a return to traditional diets, organic farming and consequently a more varied diet for more people. It's a preventive approach to the rural poverty, to the endemically low immunity of so many Indians, and consequently to the diseases facing his country. 'By a return to traditional farming, foods and lifestyles, the economy and health of our people will improve,' he says.

India doesn't need biotechnology is the message. 'Monsanto cannot feed India,' Chaudhary argues. 'What we need instead is a return to traditional farming practices, a return to biodiversity-based ecological farming which is low cost, conservation oriented, low input-high yield system as affirmed by UN-IAASTD. It is the most appropriate technology.' Farmers have reacted angrily to the problematic forcible introduction of agricultural biotechnology in India.

Chaudhary illustrates with an example from one region in southern India, where in only one month just under 500 farmers committed suicide. He explains why: 'The biotech salesmen told the farmers that their profits would skyrocket with these new seeds, that they would need to spend far less on pesticides and that bumper crops would be a certainty'. So the farmers bought GM seed, signed loans to do so and bought pesticides to spray on the buffer zones. But agriculture, being the messy, unpredictable business that it is, didn't play nice. The harvest was one of the region's worst, with monsoons wiping out every crop and the farmers left with nothing but debt.

Chaudhary takes over the narrative, explaining that the shame of debt was too great for many of these farmers: 'Some killed themselves, some sold their kidneys or their children's organs just to stay alive. I'm not making this up; these are the results of industrial agriculture'. Since then, we have been fighting industrialized agriculture on every front. As Chaudhary says, 'We're fighting against GM crops, water privatization, land acquisition, unreasonable regulations and the government'. Our organization is also working directly with the farmers, holding workshops to explain their rights, how the biotech system works and how they can regain control over their seeds and land.

'Don't think we're blindly against science or technology,' Chaudhary says, 'we're right behind anything that can help the farmers. However, almost every instance where GM crops have been introduced in India, the farmers have suffered terribly'. As Chaudhary explains, 'By its very nature, biotechnology must inherently change agricultural systems'. Farmers from America's Mid West to India's Deep South have always shared, cleaned and re-sown their own seeds. But with biotechnology, wherever it may establish itself, seeds become patented commodities, farmers sign complicated legal contracts with every purchase of GM seeds and this binds them to an agreement not to re-sow.

This means that farmers are obliged to buy new seeds every year. Companies such as Monsanto have gone so far as to set up toll-free phone numbers where people can phone in and report neighbours, friends or family suspected to be re-sowing their seeds. Chaudhary says, 'This monopoly system means that the farmers become slaves'. Our 'diversity is the basis of our rural future. Our farming culture is based on the tradition of sharing seeds, knowledge and benefits. Industrialized, GMO based agriculture has no room for sharing.'

'But this diversity is under great threat, from the globalization taking over our agro-economic system and culture,' Chaudhary continues, 'Kellogg's and Nestlé produce nearly all the food sold in our country today. Our national foods don't exist anymore in many parts of India and because of this our agricultural systems are under threat. Monocultures are eating away at small scale, patchwork agriculture.' " Chaudhary nods and concludes that, 'Biodiversity means traditional foods, traditional health, satisfying livelihoods and no more urban poverty. We must keep poverty off the streets and, with a healthy agriculture system, this would happen."

Source: www.slowfooduk.net; July, 15, 2002

78

India should set up
Power Alcohol Production

Ghaziabad, May 29, (UNI) – India should step up production of power alcohol to reduce 20% of its petrol consumption, thereby saving considerable amount of foreign exchange on import of petroleum products, says a sugar-cane expert.

The new auto fuel "gasohol" a blend of gasoline (petrol) and ethanol, in the ratio of 80:20 would not only be cheaper but would also check pollution, Indian Sugarcane Development Council Chairman Dr. Krishan Bir Chaudhary told UNI.

He said Brazil, the world's largest sugarcane growing country, was already producing power alcohol successfully. India, which ranked second after Brazil in the production of sugarcane, could also follow suit he added.

Dr. Chaudhary said he had sought from Brazil the technological know-how for the production of power-alcohol, popularly known as "Proalcool" in that country and "they were keen to supply information to India". The Brazilian authorities had already supplied to India the preliminary information about the technology he said.

Sugarcane in Brazil is largely grown as a source of alcohol and more than a million motor vehicles run on this fuel.

Dr. Chaudhary a power-alcohol expert besides being an authority on sugarcane production said experiments had revealed that power-alcohol-run vehicles caused less air pollution compared to vehicles consuming pure petrol.

Toxicity of fumes exhausted by alcohol and gasoline-powered vehicles was found to be very low. In fact gasoline-fuelled vehicles produced more Carbon Monoxide (CO), Sulphur Dioxide (SO2), which caused damage to human cells, and Hydrocarbons (HC), which could cause Cancer, he pointed out.

Dr. Krishan Bir Chaudhary, Chairman, Indian Sugarcane Development Council, inaugurating the 16th council meeting

However, the alcohol-engines emit more Aldehydes, but the level of emission is, by no way, dangerous.

Dr. Chaudhary, quoting a Brazilian report, said the use of alcohol-engines had reduced the level of lead particulates in the atmosphere by 75% in the city of Sao Paulo as an alcohol engine used alcohol- removing lead-based additives.

If the power-alcohol is manufactured year-round from cane molasses, the additional equipment needed in the existing distilleries are: dehydration columns, separate boiler and turbo alternator and cooling tower with a total additional capital investment of Rs.175 lakh.

The cost of production will be Rs. 7.40 per liter and Rs.5.83. per liter for 30,000 liter per day capacity of power alcohol plant working for 180 and 300 days in a year respectively.

Dr. Chaudhary said the power-alcohol, if produced in India would reduce the annual petroleum consumption of the country, currently at 58 million tons, by 20% and total expanding would come down by about 10%. The growth of petroleum consumption was at 9.4% during 1990-91.

As such the new device would further consolidate savings of petroleum and amount spent on its import. According to tentative estimates, All India deficit of petroleum products met by imports was 25 million tones in 1991-92.

Furnishing more data, Dr. Chaudhary said the installed annual capacity of 200-old distilleries in India was 15,86,928 Kilolitres. While actual production in 1988-89 was 797.32 million liters. As the total requirement of alcohol by industries in 1988-89 was just 545 million liters, it was possible to obtain surplus alcohol for fuel purposes, he noted.

He said sugarcane juice was also being used for the production of power alcohol in Brazil, where about 58 liters of alcohol was produced from one tone of sugarcane.

Source : Patriot & National Herald; May, 29,1992

Author on vanguard
of Farmers' Interests

79

Wipe out Draught Tears

15. April, 2013

My dear Sh. Sharad Pawar ji,

Coming to know about the problems faced by the farmers in Maharashtra on account of drought, I visited the State on April 6 to 10, 2013. My visits were to the interiors of Buldhana, Jalna, Aurangabad, Beed, Ahmednagar districts.

I came to know the sufferings of the farmers which are really shocking. Apart from the farmlands being parched, I found dead trees, dry wells, dry bawris and dry rivers in Buldhana, Jalna and Beed. Practically most parts of Marathwada, western Maharashtra and parts of Vidarbha are left dry.

The State Government is unable to supply drinking water by tankers. The State administration may on records be showing that they are supplying drinking water, but during my personal visit I found no tankers supplying drinking water in villages.

It is a lesson for us to encourage water harvesting on a war footing to recharge ground water table.

Social organizations have set up relief camps for animals (Chawnis) and the State government is rendering help which is not enough to meet the requirements of fodder, feed and drinking water for animals. I am enclosing some representations received by me on visit to these animals relief camps for your kind perusal.

From the impressions I have gathered on my visit about the sorrowful tales of farmers, I would suggest that all types of loans along with interests of farmers in these affected areas should be completely waived off and fresh loans extended to farmers at zero per cent interest rate repayable by them at a longer period.

With kind regards

Yours Sincerely

To,   (Krishan Bir Chaudhary)

Sh. Sharad Pawar
Minister of Agriculture
Krishi Bhavan
New Delhi

80

FDI In Retail to Kill Livelihood

01 Oct. 2012

My dear Sh. Anand Sharma ji,

As per our discussion, we are submitting the charter of demands with respect to FDI in multi brand retail for your kind consideration to protect the farmers interest of the country.

1. Establish a National Authority to act as a watchdog with majority representation of farmers. This authority needs to have the power and resources to directly intervene and take action against anti-supplier practices without requiring a complaint. Farmers should be able to the authority for free.

2. Require all corporations to reveal their supply sources and buying prices. Maximum Retail Price (MRP) must not be left open for retailers and there needs to be a reasonable cap on MRP in proportion to the input cost. Ensure minimum 60% share of retail price to producers of Milk, Fruits and Vegetables.

3. Regulate and monitor contact farming to protect the interests and land of farmers. Payment to farmers should be within a month of procurement and no unreasonable rejection of farmers produce in the name of quality and standard.

4. Ensure that no single retailer monopolizes procurement operations in an area. Prohibit vertical agreements between retailers or intermediaries and seed and fertilizer companies.

5. Food retailers or other agribusiness companies should not be allowed to corner and hoard food-grains stocks under any circumstances. To prevent cornering of stocks by corporate, there should be rules for public disclosure of stock holding levels. Public agencies should be empowered to purchase food-grains from the private holders at pre-specified prices.

We hope that you will take the necessary action in the best interest of the farming community of India.

With kind regards,

To,   (Krishan Bir Chaudhary)

Sh. Anand Sharma
Union Minister of Commerce & Textile,
Udyog Bhawan,
New Delhi

81

Save Farmers' Land from Corporate Loot

28.March.2012

To,

The Chairperson
Parliament standing committee on Rural Development
Parliament House NXE, New Delhi

Respected Madam,

Bharatiya Krishak Samaj on behalf of farming community request you to consider the following changes in the Land Acquisition Amendment Bill that deserve inclusion in the Bill :

Public Purpose definition should be limited to core functions of the government performed with the public money and in no case acquisition should be made for the private corporations where they get benefits due to any forcible acquisition of land or any other natural wealth under this act. Any project drawing private profit can't be considered public purpose.

Section 26 read with Schedule 1 freezes compensation to twice the registered or stamp value in rural areas. In other words, landowners cannot ask for more. If this is the case, where is the room for negotiations before they give their consent? The section should clearly state that twice the registered or stamp value is the minimum, and it could be more depending upon the agreement with the landowners. Similarly all RR benefits as given in the Schedule should be flexible, subject to the minimum prescribed.

Compensation of land should be based on the converted or future land use. It should be based on highest sale for similar lands in adjacent areas, multiplied by a factor of Six times (including solatium) in rural areas and four times (including solatium) in urban areas.

The 10% developed land should be given to Landowners without any charge. The cash amounts mentioned in the bill should all be linked with inflation.

Section 28A of the old Act provided for benefits to All affected families of the enhanced compensation awarded by the Court to one individual. There is no such provision in the current Bill. One should have a similar section in the new law too.

The farmer/landowner should also be permitted to do the same activity of development as the authority permits to the builders on the same conditions and on payment of external development charges.

From section 3(za)(iv) the phrase 'planned development' to be removed as it is quite vague and is prone to be misused. Even acquiring land for companies and builders can be justified in the name of 'planned development', and then they will not be subject to the 80 percent consent clause.

Land that is not used within five years should be returned to the owner. Section 95 which states that it shall be transferred to the State Government's Land Bank in fact contradicts section 93 which states that 'No change from the purpose or related purposes for which the land is originally sought to be acquired shall be allowed'.

From section 96 the phrase, 'without any development having taken place on such land' may be deleted, as it completely defeats the intention behind this section of sharing capital gains with the landowner.

Chapter XI on 'TEMPORARY OCCUPATION OF LAND' should be deleted. This allows government to acquire land for a company against farmers' wishes. Companies are likely to misuse the powers given under this section, as they will first take land on a temporary basis, and when it becomes totally unfit for cultivation then ask the Collector to acquire it permanently. No arbitrary powers should be given to the government as proposed in chapter XI of the bill.

In section 3(c)(iv) the word 'acquisition' may be replaced by 'acquisition and resumption' as forest lands and water bodies are not acquired, these are just resumed. In the absence of this amendment users of common land & forests and slum dwellers will get nothing, not even RR benefits. Thus the poorest people will be deprived of their livelihoods without any RR benefits. This amendment is particularly relevant in view of the state governments' reluctance to implement the community clauses of FRA, such as Section 3(1)(b) to 3(1)(m) of FRA.

Serial number 1 in the second Schedule should clearly guarantee a separate house to each affected family. The present draft only promises house for house, even when two or three families are living together in one house. This defeats the very purpose of defining families as nuclear families, and not joint families.

To save the food security of the country any transfer of three crops agricultural land to non-agriculturists, in general, and to foreigners and NRI's, in particular, be prohibited immediately. We hope that you will consider these suggestions and make changes in the Bill.

With kind regards,

(Krishan Bir Chaudhary)

82

Farmers ignored once again

14th March, 2007

Respected Smt. Sonia Gandhi Ji,

The International Seminar on "Saving Doha and Delivering on Development" was organised by the Ministry of Commerce in collaboration with Pro-WTO Non Governmental Oraganisations on 12th & 13th March, 2007 in New Delhi. It's surprising that no farmers' organization was invited in the seminar to raise the farmer's issues. These NGO's are working on the directions of the foreign funding agencies in their interests and are known for Pro-WTO.

The WTO agriculture negotiations have made little progress so far. Several deadlines have been missed due to undue stubbornness by the EU and US. Both the rich blocks are seem not willing to understand that the Doha round was declared a 'Development round' and not a market access round. They are also not willing to comprehend the importance of agriculture for poor countries.

Therefore, agriculture remains the main stumbling block in this round of negotiations too. While the EU and US are reluctant to reduce their farm subsidies and the level of protection to their farm sector, they have been united in pushing poor countries to open up for more market access.

However, this time the situation is different from the Uruguay round of negotiations. Developing countries are well prepared and they have organised themselves to safeguard their interests. The two important blocks of developing countries are the 'G 20' and 'G 33'. The G 20 is very active on domestic farm subsidies and tariff issues, and the G 33 equally stood well by pushing forward developing countries' interests to protect agriculture through protective proposals, known as 'special products' and the 'special safeguard mechanism'

In the Uruguay Round Agreement on Agriculture, many countries maintained the right to defend their agriculture sector by using 'special safeguards' available under article 5.1 and developed countries specially benefited from the 'peace clause' - which exempted them from having their cheap exports challenged legally at the WTO.

As part of the Doha round of WTO negotiations on agriculture it has been agreed - while final modalities and measures are yet to be decided - that developing countries would have a category of agricultural products classified as 'special products', which would have lower reductions from general formula cuts.

Similarly a 'special safeguard mechanism' which allows developing countries to increase their import tariff levels in case of import surges from abroad. The objective of these measures is to ensure food and livelihoods security and rural development in developing countries by protecting small farmers against the volatility of the world prices.

The 2004 July Framework in the Doha talks provided developing countries flexibility to designate certain agricultural products as special products based on the criteria of food security, livelihood concerns and rural development in order to build in flexible treatment. The Hong Kong Ministerial reaffirmed that developing countries had a right to self-designate an appropriate number of tariff lines on agricultural products as special products and to develop special safeguard mechanisms to protect poor farmers from import surges. However since then, these measures have faced strong attacks from many developed countries and institutions who seem to want to increased market access for farm goods at any price.

Contrary to the resistance from rich countries and international financial institutions, we cannot leave millions poor farmers at the mercy of global market forces. Already liberalisation of agriculture sector under structural adjustment programmes and the Uruguay round of commitments have devastated millions of farmers throughout the developing world.

The losses are immeasurable. Rural unemployment has increased because subsidised cheap agricultural imports have flooded in India. Indian farmers are not getting the remunerative prices of their produces. More than one lakh farmers have committed suicides due to these policies. When India export agricultural produces the global prices fall and when India imports agricultural produces gradually the prices rise-up. This has been the case when India imported Wheat last year.

Therefore in order to protect local employment and protect our food sovereignty, our government need to have an offensive approach in negotiating for enhanced special products and special safeguard mechanisms. By no means should we surrender to the pressure of the western countries or international financial institutions.

Finally, to restructure the future discourse of trade negotiations and making them favourable to the needs of developing countries and offer appropriate 'policy space' that allows developing countries to pursue agricultural policies supportive of their development goals, poverty reduction strategies, food security and livelihoods concerns, we urge you that you give the directions to Mr. Kamal Nath (Union Commerce Minister) to participate in the G 33 meeting in Jakarta on 20th March, 2007 to strengthen the case of poor small farmers.

With kind regards,

(Krishan Bir Chaudhary)

To,

Smt. Sonia Gandhi
Chairperson UPA
10, Janpath, New Delhi

83

Wheat Import will damage Food Security

19.June.2006

Dear Sh. Sharad Pawar Ji,

We are very much concerned that under the recent relaxed quarantine norms for wheat imports there is a possibility of at least 32 exotic weeds entering in the country. Also there are chances of imported wheat containing residues of over 60 pesticides which may be cause health hazard to our citizens. The Govt. of India should make public all the processes and tests relating to risk analysis, including those relating to quarantine and sanitary and phytosanitary (SPS) measures. Imports of all food and agro produces should be cleared only after the consignments are vouched for their safety through adequate quarantine and sanitary and phytosanitary (SPS) tests.

Our sources have revealed that ministries dealing with the import of wheat are forcing scientists to give favourable reports, much against their will. Their only intention is to facilitate entry of sub-standard wheat from US and Australia. In fact the series of moves by the government for wheat imports has invited public suspicion. Wheat is being imported when there is a good harvest of 73.06 million tonne (estimate arrived at on basis of reports from producing states). This estimate was deliberately scaled down to 71.54 million tonne, just a day before the Parliament was to reconvene after the recess. It was done with the intention to justify imports and escape the wrath of the Opposition in Parliament. Next area of suspicion is in the government has incrementally increased to quantity of wheat to be imported. In Feburary, 2006, government decided to import 500,000 tonne wheat. The USDA said that this quantity of imports would not be enough. India needs to import extra 3 million tonne wheat. The Indian government then decided to import additional 3 million tonne. Now USDA says that India's total import requirement should be 4.5 million tonne that means an additional import of one million tonne. The Indian government has also responded to this and has indicated its intention to complete the import of 4.5 million tonne wheat. What does this show? Are we relying on our own production data or working at the behest of USDA ?

The tender floated on May 8, 2006 for import of 3 million tonne wheat resulted in finalisation of contracts for import of only 800,000 tonne. Most of the bidders could not meet even the relaxed quality and quarantine norms specified in the tender. Government then thought of further relaxing the quality norms and floated a tender on June 12, 2006 for import of the remaining 2.2 million tonne wheat. Scientist are saying there are serious quarantine problems of wheat from Australia and US but they are pressurized to give favorable reports. According to agricultural scientists there can be serious problems for Indian farmers because the import of wheat from Australia and US will add new infectious diseases and new types of weeds in India.

The Government should, therefore, disclose all the scientific facts in public domain. We want to know the details of process employed in analysis and test results. The Farmer's Organizations will set-up a independent panel of scientists to evaluate the Govt.'s test results, including those relating to quarantine and sanitary and phytosanitary (SPS) measures.

With kind regards,

To,   (Krishan Bir Chaudhary)

Sh. Sharad Pawar
Minister of Agriculture
Krishi Bhawan, New Delhi

84

Illegal Patent on Indian Wheat Revoked

8th October, 2004

Respected Prime Minister,

Sub:- Illegal patent on Indian Wheat Revoked

You will be happy to know that we have taken initiative to fight for Indian farmers to save Indian Wheat variety which was patented in European Patent Office in Germany.

I myself Mr. Krishan Bir Chaudhary on behalf of Krishak Samaj along with Greenpeace had filed opposition in Germany on 27th January, 2004. The patent on an Indian Wheat variety (EP445929) was granted on May 21st, 2003 to the multinational corporation Monsanto.

Monsanto got this patent by transferring genes from an Indian wheat landrace (NAPHAL) to their variety. They claim this act of theft as an " Invention". This patent was a blatant example of Bio-piracy as it forfeits the rights of Indian farmers and dismisses years of hard field work.

The Indian farmers was deeply outraged by this attempt to monopolise Indian Agriculture. We immediately take up this issue with EPO and ask them to revoke this illegal patent.

This is to bring your kind notice that patent on Indian Wheat variety is revoked in total by the European Patent Office, Munich

(Germany). This victory will go a long way to save the rights of the farmers in India.

With kind regards.

Yours sincerely,

(Krishan Bir Chaudhary)

To,

Dr. Manmohan Singh
Prime Minister of India
New Delhi

85

PM, please lend your ears to Farmers' Woes

2 8th June, 2004

Respected Dr. Manmohan Singh ji,

Kindly accept our thanks for the invitation to attend the meeting. I have highlighted here under a few vital points and I am sure they will form basis for a realistic discussion on agriculture.

1.　　**Minimum Support Price.** -To protect the food security of the country and ensuring remunerative prices to the farmers, the Govt. should continue to give minimum support price for all agriculture produces. The major responsibility in this respect lies with the Central Govt. and the Central Govt. should co-ordinate efforts with the State Govt. to procure the farm produces from the farmers. Only the remunerative prices will desist the farmers from committing suicide.

2.　　**Subsidy.** - The Govt. should give subsidies directly to farmers to sustain agriculture otherwise Indian agriculture will be ruined. The quantum of subsidies in developed countries is much higher in comparison to farm subsidies given in India. In view of the stiff global competition, the Central Govt. should substantially enhance subsidies on a wide range of agriculture inputs such as seeds, fertilizers, power and irrigation etc.

3.　　**Quantitative restrictions (Q Rs).** - With wavering

any further, the Central Govt. should take immediate steps to impose Q Rs otherwise the agriculture sector will be ruined. The country has already become a dumping ground of unwanted imports.

4. **Tariff.** - The Central Govt. should impose high tariff on imports of unwanted agriculture commodities especially edible oils, sugar etc. to protect agriculture.

5. **Irrigational Facilities.** - The Govt. should implement various time bound irrigational projects, promote Watershed management programme. Country wide campaign should be launched for the promotion of Drip and Sprinkler irrigation to save the water. The system should be exempted from excise and all taxes & should be made easily available through subsidy.

6. **Credit & Interest.** - The Govt. has to stop discrimination to the farmers vis-à-vis the elite business class of India. Automobile and Home loans are available on cheaper interest rate. Therefore, if the Govt. has to save agriculture it should ensure easy flow of credit to the farmers on minimal possible interest.

7. **Marketing of Agricultural Commodities.** - Marketing of Agricultural Commodities needs a new approach and strategy. To avoid post harvest glut of Agricultural Commodities and ensure remunerative prices to the growers, the management of regulated Agro-market should be so streamlined that instead of corporate and bureaucratic control, it should be manned by the farmers because they are directly involved with production. There should be direct linkage between production, storage, marketing and distribution. The marketing pattern developed on this line will be very effective in utilization of rural manpower and will give employment to the rural youths.

8. **Post Harvest Technology.** - To save the perishable agricultural commodities from rot and avoidable storage problem, right infrastructure has to be created for the development of Post Harvest Technology. Fruit and Vegetable processing, canning packaging, refrigerated transportation, quarantine certification must be fully explored to expand domestic and overseas markets.

9.   Seed. - Seed village pattern should be developed at block level for easy availability of quality seeds to the growers. The seed public sector should be modernized and strengthened. Seeds being the basic input, it will ensure food security and at the same time it will protect our farmers from being exploited by MNCs.

10.   Sugarcane complexes. - Sugarcane though being grown extensively in BRAZIL, various products derived from Sugarcane other than sugar like Ethyl alcohol, Citric acid, Lactic, Cattle feed, Ephedrine hydrochloride, Oxalic acid, Baker's yeast, Monosodium glutamate, Lysine, Acetone-butanol alcohol, Acetic acid, Acetic-anhydride, Polyethylene, paper and Synthetic rubber. Another important by-product of Sugarcane is Bagasse which is used for manufacture of hard boards, plastic, gums etc. These products have become commercially more viable due to their industrial utility. Not only the economics of Sugarcane cultivation and recovery of sucrose should be taken into account but the commercial and industrial aspect of Sugarcane must be considered while fixing SMP of Sugarcane. Profits gained from these by-products of Sugarcane go to the coffer of sugar mills. But the poor Sugarcane growers are paid so inadequately that they can't even meet the cost of production. For the low recovery, the sugarcane growers can't be held responsible. Since they are the producer of the basic raw material, they have the right to share the profit and gains from the industrial by-products of sugarcane. Therefore, it is imperative to develop Sugarcane complexes on BRAZIL pattern.

11.   Organic farming. - The enormous infusion of synthetic plant nutrients and application of agro-chemicals have severely damaged and disturbed the ecological balance of the soil. To save the soil from further deterioration, the only alternative available is Organic farming. Organic farming has been accepted as the highly sustainable farming system as it maintains the physical structure and fertility of the soil. Soil organism play important role in recycling plant nutrients and provide the plants with useful growth hormones. Organic farming does not cause any degradation of soil, degeneration of soil microbial diversity and aids in the preservation and multiplication of soil microbial flora. The system of organic farming involves eco-friendly technology that take care of small land holdings financial constraints of poor farmers. It is therefore, highly imperative to explore and harness the potentiality of organic farming for ensuring perpetual food security and maintaining the fertility status of the soil.

12.   Transgenic seed and G.M. Food. - The Central Govt. should put a complete freeze on import of transgenic seeds and G.M. foods and should ban research on this technology. The Gene Giants are bet upon introducing the dreadful Terminator gene technology into the country. Terminator gene itself and the resulting cross pollination will produce only sterile seeds incapable of further multiplications and our food security will be endangered.

13.   Crop Insurance. - Only the formulation of a comprehensive crop insurance scheme will not serve the real purpose unless it is implemented in latter and sprit.

14.   Dairy, Poultry & Fisheries. - Dairy, Poultry & Fisheries can't be considered into isolation with the overall development of agriculture because they are capable of generating employment & influence rural economics.

15.   North East. - Special agriculture policy to be framed for the north east region taking into account the Geo-climatic conditions and with special emphases on rich floral diversity and abundance of herbal and Medicinal plants.

16.   The Government should immediately remove the Capital Gain Tax on compensation given to farmers, of their land acquired by the Government.

With kind regards.

Yours sincerely,

(Krishan Bir Chaudhary)

To,

Dr. Manmohan Singh
Prime Minister of India

शरद पवार
SHARAD PAWAR

D.O. No. 6/8 /AM
कृषि एवं खाद्य प्रसंस्करण उद्योग मंत्री
भारत सरकार
MINISTER OF AGRICULTURE &
FOOD PROCESSING INDUSTRIES
GOVERNMENT OF INDIA

DO No.13035/46/2012-PP.I
/ 8 February, 2013

Dear Dr. Chaudhary,

Kindly refer to your letter dated 24.09.2012 requesting for inquiry into malpractices in granting registration of Imazethapyr Technical to Makhteshim-Agan India Private Limited.

I have been informed that notice has been issued to M/s. Makhteshim-Agan India Private Limited under Section 11 of the Insecticides Act, 1968 to explain why certificate of registration granted to them for Imazethapyr Technical should not be revoked. The matter has also been brought to the notice of Vice Chancellor, Bidhan Chandra Krishi Vishwavidyalaya Kalyani, West Bengal for appropriate action.

With regards,

Yours sincerely,

(Sharad Pawar)

Dr. Krishan Bir Chaudhary,
President,
Bharatiya Krishak Samaj,
F-1/A, Pandav Nagar,
Delhi-110091.

Office : Room No. 120, Krishi Bhawan, New Delhi-110 001 Tel.: 23383370, 23782691 Fax : 23384129
Resi. : 6, Janpath, New Delhi-110 011 (India) Tel. : 011-23018870, 23018619 Fax : 011-23018609
E-mail : sharadpawar.sp@gmail.com

RAJIV MISHRA
CHIEF EXECUTIVE
LOK SABHA TELEVISION
LOK SABHA SECRETARIAT

23, MAHADEV ROAD
NEW DELHI-110001
Tel.: 011-23739590, 23738965
Fax: 011-23739641
E-mail:rajiv.mishra@sansad.nic.in

D.O.No.CE/Thanks/2012-13

Dt: 08.01.2013

Shri Krishan Bir Chaudhary,
President, B.K.S.
F-1/A, Pandav Nagar,
New Delhi-110091.

Dear Shri Chaudhary ji,

At the outset, I would like to thank you for participating in Lok Sabha Television (LSTV) as one of the esteemed guest. Your inputs were really very valuable, beneficial, worthy and useful for our viewers as well as for the host/anchor.

As your good self is aware, along with parliament, the LSTV shares a responsibility to contribute to political, economic and social dvevelopment in ways consistent with democratic principles by pursuing fact-based, full substantiated coverage and to inform citizens about current affairs, parliamentary activities and the role of parliament so as to help empower the public.

I would like to add that you have very clear and good understanding of the facts and issues related with your domain, and you do have the ability to put them together in a very coherent and convincing manner.

Thank you very much for participating in LSTV.

With Warm regards,

(RAJIV MISHRA)

Parliament Office: F004, Parliament Library Building, New Delhi-110001

United States
Department of
Agriculture

Office of
Agricultural
Affairs

American
Embassy

Shanti Path,
Chanakya Puri
New Delhi
110 021
INDIA

Telephone:
+ 91-11-
2419-8000

Facsimile
+ 91-11-
2419-8530

February 29, 2012

Dr. Krishan Bir Chaudhary
President
Bhartiya Krishak Samaj
F-1/A, Pandav Nagar
Delhi 110 091
Telefax: 0011-22751281; +91-98103-31366 (cell)
Email: kbc@kisankiawaaz.org

Dear Dr. Chaudhary:

I am the new Minister-Counselor for Agricultural Affairs at the Embassy of the United States of America.

I respectfully request a meeting with you to introduce myself and to discuss the U.S.-India agricultural relationship.

Dr. Santosh K Singh, Agricultural Specialist, from my office will accompany me to this meeting. If you consent, my secretary will follow up with your office to set up a convenient time.

Sincerely,

Allan Mustard
Minister-Counselor for Agricultural Affairs

Dr. Krishan Bir Chaudhary, & Mr. Allan Mustard Minister-Counselor for Agricultural Affairs, American Embassy New Delhi, discussing agricultural issues in BKS Office on 23.April.2012 .

SONIA GANDHI
CHAIRPERSON
NATIONAL ADVISORY COUNCIL

2, MOTI LAL NEHRU PLACE,
NEW DELHI-110 011
PHONES : 011-23062596
23062582
FAX : 011-23062599

NO. Z-11011/100/2011-NAC-2621

12 July, 2011

Dear Dr. Chaudhary,

I have received your letter dated 5th July, 2011 regarding large scale sale of sub-standard pesticides/insecticides. You have pointed out that no action was taken against the Pesticide Mafias inspite of bringing the matter to the notice of the concerned officials.

My office is taking up the matter with concerned authority for appropriate action.

With good wishes,

Yours sincerely,

Dr. Krishan Bir Chaudhary,
President,
Bharatiya Krishak Samaj,
F-1/A, Pandav Nagar,
Delhi-110091.

Phone : 23019080

ALL INDIA CONGRESS COMMITTEE
24, AKBAR ROAD, NEW DELHI - 110 011

Sonia Gandhi
President

June 19, 2010

Dear Dr Chaudhary,

I have received your letter of 16th June 2010 inviting me to an International Convention on "Impact of Global Climate on Agriculture", jointly organized by Bharatiya Krishak Samaj and the Russian Centre of Science and Culture on 28th July in New Delhi. I regret my inability to accede to your request owing to pressing engagements during that period, however, I send my good wishes to the organisers for the success of the programme.

With good wishes,

Yours sincerely,

Dr Krishan Bir Chaudhary
President
Bharatiya Krishak Samaj
F-1/A, Pandav Nagar
Delhi - 110 091

डा० एम. वीरप्पा मोइली
Dr. M. VEERAPPA MOILY

मंत्री
विधि एवं न्याय
भारत सरकार
402, 'A' विंग, शास्त्री भवन,
डा. राजेन्द्र प्रसाद रोड,
नई दिल्ली–110 115
MINISTER OF LAW & JUSTICE
GOVERNMENT OF INDIA
402, 'A' WING, SHASTRI BHAWAN,
Dr. RAJENDRA PRASAD ROAD
NEW DELHI-110 115

July 26, 2010.

Dear Dr. Chaudhary,

I am glad to know that an International Convention on "Impact of Global Climate on Agriculture" is being organized jointly by Bharatiya Krishak Samaj and the Russian Centre of Science and Culture on 28th July, 2010.

With the rapid pace of development of the human race, we have been exploiting natural resources to such an extent that the nature's cycle and the system of checks and balances have started getting disturbed. Rapid climate change has prompted serious concern over the potential consequences of global warming to the world's ecological systems and agriculture. Burning of fossil fuels has altered the delicate balance of earth's environment and the effects are already being seen. We can still do a great deal to avert the more serious consequences of climate change by making rapid changes in the ways that we make and use energy, consume, travel and communicate.

The theme of the Convention has been very aptly chosen and I am sure the deliberations in the Convention will go a long way in suggesting measures to meet the challenges.

I send my greetings to the organizers as well as the participants and wish the proposed Convention a grand success.

(Dr. M. Veerappa Moily)

Dr. Krishan Bir Chaudhary,
President,
Bharatiya Krishak Samaj,
F-1/A, Pandav Nagar,
Delhi – 110 091.

Lt Gen (Retd.) MM Lakhera
PVSM, AVSM, VSM

RAJ BHAVAN,
AIZAWL - 796 001
Phone : 0389 2322262 / 2323200
E-mail : rbaizawl@sanchamet.in
Fax : 0389 2323344

DO No. F.23015/3/2008-GS
5th April 2010

Dear Dr Krishan Bir Chaudhary,

It was a pleasure meeting you at Delhi during my recent visit. At the outset I would like to compliment you on your efforts to protect the interest of Indian farmers. There is no doubt that we need to increase the productivity of our farmers. This will help to improve their economic condition as also make the country self-sufficient in all types of food grains. However this has to be done taking the Indian environment into account.

I am enclosing a copy of Biotech Newsmagazine for your information. You will note that on page 14 to 21 there is an article on Bt Brinjals, wherein a number of scientists have been quoted. What is surprising that all across the board have supported introduction not only of Bt Brinjals but also of other GM crops. It is evident that this is a one-sided story, however such stories do carry a message. I think there is a need to publish the views of the other side also so that a balanced picture is projected.

I thought I would send this to you for your information and action deemed fit.

With best wishes and regards,

Lt. Gen. M. M. LAKHERA
PVSM, AVSM, VSM (Retd)

Dr Krishan Bir Chaudhary
President
Bharatiya Krishak Samaj
F-1/A Pandev Nagar
Delhi 110 091

डा.टी.रामसामी
सचिव
Dr. T. RAMASAMI
SECRETARY

भारत सरकार
विज्ञान और प्रौद्योगिकी मंत्रालय
विज्ञान और प्रौद्योगिकी विभाग
टेक्नोलाजी भवन, नया महरौली मार्ग, नई दिल्ली-110 016
GOVERNMENT OF INDIA
MINISTRY OF SCIENCE & TECHNOLOGY
DEPARTMENT OF SCIENCE & TECHNOLOGY
Technology Bhavan, New Mehrauli Road, New Delhi-110 016

D.O. No. DST/Secy/ 90 /2010   24th February, 2010

Dear Dr. Chaudhary,

I receive your letter of 5th February, 2010 along with a copy of the resolution passed in National Convention of Bharatiya Krishak Samaj held on 27th December, 2009.

Majority of the resolutions pertain to trade and industrial infrastructure aspects of agri-business.

I have directed my colleagues in the Ministry of Science and Technology to address issues wherever technology and trade are coupled.

Thank you for the valuable feedback.

Kindest regards,

Yours sincerely,

(T. RAMASAMI)

Dr. Krishan Bir Chaudhary
President,
Bharatiya Krishak Samaj,
F-1/A, Pandav Nagar,
Delhi – 110 091

Greenpeace e.V · 22745 Hamburg

Krishan Bir Chaudhary
Pass port No. E 3769827
A-1, Nizamuddin West
New Delhi -110013
India

contact person:
Christoph Then
phone: +49(0)40/30618-395
fax: +49(0)40/30631-195
christoph.then@greenpeace.de

Hamburg, february 21th, 2007

Dear Mr. Krishan Bir Chaudhary,

I am pleased to invite You to attend our press conference and following events on the patenting of seeds in March 26th - 30th, 2007 in Munich and Berlin, Germany.

We are sending your air ticket and we have booked accomodation in hotels in Munich and Berlin. The all expenses of your visit will be borne by us.

We are looking forward to meet You!

Yours sincerely

Dr. Christoph Then
Department Genetic Engineering, Agriculture, Toxics

Dr PM Bhargava
Vice Chairman

National Knowledge Commission
Government of India

9th February 2007

My dear Krishan Bir ji,

I was absolutely delighted to spend some time with you in Delhi on the 6th February evening. I am enclosing copies of the four reports of the meetings organized by the ICAR on the suggestion of the NKC, one each on Post-Harvest Technologies, Organic Farming, Integrated Pest Management and Biopestides and Energy Use in Agriculture.

I am also enclosing a list of 21 areas that relate to agriculture where existing knowledge needs to be conveyed to the farmers and used, and/or new knowledge needs to be generated for the same purpose. The four areas mentioned above are out of these 21 areas. ICAR is planning to hold meetings in the other areas as well and I believe we should co-operate with them.

I have no doubt you will like the contents of the four reports I am sending which have been approved by the ICAR. They are very keen to implement these reports if they come back to them after approval of the National Knowledge Commission or the Government.

With warm personal regards,

Yours sincerely,

P M Bhargava

Dr Krishan Bir Chaudhary
A-1, Nizamuddin West
New Delhi 110 013

"Furqan Cottage" 12-13-100, Lane # 1, Street # 3, Tarnaka, Hyderabad - 500 017

Tel : (Off) +91-40-2701 7789, 2701 5089, 2700 5504 ; Fax : +91-40-2701 7857 ; E-mail : bhargava.pm@gmail.com / pmb1928@yahoo.co.in

Phone : 23019080

ALL INDIA CONGRESS COMMITTEE
24, AKBAR ROAD, NEW DELHI - 110 011

Sonia Gandhi
President

June 8, 2006

Dear Shri Chaudhary,

I have received your letter dated 31st May, 2006 alongwith enclosures regarding the decision of the Government to Import Wheat and Sugar.

My office is requesting the concerned Ministry to look into the issues raised by you.

With good wishes,

Yours, sincerely,

Dr Krishan Bir Chaudhary,
A-1, Nizamuddin West,
New Delhi – 110 013

Prof. M. S. Swaminathan
Chairman

राष्ट्रीय किसान आयोग
भारत सरकार
कृषि मंत्रालय
(कृषि एवं सहकारिता विभाग)
NATIONAL COMMISSION ON FARMERS
GOVERNMENT OF INDIA
MINISTRY OF AGRICULTURE
(DEPARTMENT OF AGRICULTURE & COOPERATION)

February 27, 2006

Dear Dr. Krishan Bir Chaudharyji,

Thank you very much for your kind letter concerning the Indo-U.S. agreement on agricultural research. Dr. Mangala Rai had asked me to address the Group briefly and I mentioned to them how we have 25% of the global farming population in India and why the livelihood and income security of 70% of India's population engaged in agriculture should be the bottom-line of our R & D policies in agriculture. I requested Dr. Mangala Rai to send me a copy of the agreement but I give below a statement from his letter which indicates the present position:

"The institutional and financial details in respect of the programme are still being worked out. The final agreed programme will be furnished to you once firmed up."

I shall keep you informed.

With warm personal regards,

Yours sincerely,

(M.S.Swaminathan)

Dr. Krishan Bir Chaudhary,
A-1, Nizamuddin West,
New Delhi-110 013

Prof. M. S. Swaminathan
Chairman

राष्ट्रीय किसान आयोग
भारत सरकार
कृषि मंत्रालय
(कृषि एवं सहकारिता विभाग)
NATIONAL COMMISSION ON FARMERS
GOVERNMENT OF INDIA
MINISTRY OF AGRICULTURE
(DEPARTMENT OF AGRICULTURE & COOPERATION)

February 1, 2006

My dear Dr. Chaudhary,

It was a real pleasure meeting you at Contai. I thank you for the leadership you are providing for our agricultural renewal. In this context I enclose a copy of the third Report of the National Commission on Farmers. We are looking forward to your guidance and advice.

With warm personal regards,

Yours sincerely,

(M.S.Swaminathan)

Dr. Krishan Bir Chaudhary,
A-1, Nizamuddin West,
New Delhi-110 013

Encl: As above

National Bureau of Plant Genetic Resources
(Indian Council of Agricultural Research)
Pusa Campus, New Delhi - 110 012

Phone : 011 –
25783697
FAX : 091- 011 –
25842495
Resi 25071724

E. mail : directornbpgr@yahoo.co.in

Home Page :- http://nbpgr.delhi.nic.in

ICAR

Dr. J. L. Karihaloo

GEF/Nom. /No.
10th January 2006

Subject: Training Programme on "Biosafety Concerns of Transgenics and Detection of LMOs" from 16th to 20th January, 2006 - Invitation for Lecture

Dear Dr. Chaudhary,

I have the pleasure to inform you that we are organizing a training programme on **Biosafety Concerns of Transgenics and Detection of LMOs** from 16th to 20th January 2006 at NBPGR, New Delhi. NBPGR has been identified as one of the partner institute under Global Environment Facility (GEF)-World bank funded capacity building project for the implementation of Cartagena Protocol on Biosafety, coordinated by Ministry of Environment and Forests, Govt. of India. The training programme is a part of the institutional endeavor aimed primarily at developing a core group of experts at the national level in this area to tackle various emerging issues related to biosafety concerns and molecular detection of transgenes.

Your contribution and wide experience in the field is well known. We would feel privileged to have your lecture on

Transgenics and Indian Agriculture

18th January 2006 at 11.30-01.15 PM at NBPGR, New Delhi.

We would appreciate receiving a line of confirmation of your acceptance immediately. A brief note (3-4 pages) or copy of slides to be distributed during the lecture may kindly be sent in advance or you can bring along.

With kind regards,

Yours sincerely,

(J. L. Karihaloo)

Dr. Krishan Bir Chaudhary,
A-1, Nizamuddin West
New Delhi - 110013

FOOD AND	ORGANISATION	ORGANIZACION	مـنـظـمـة
AGRICULTURE	DES NATIONS	DE LAS NACIONES	الأغـذيـة
ORGANIZATION	UNIES POUR	UNIDAS PARA	والـزراعـة
OF THE	L'ALIMENTATION	LA AGRICULTURA	لـلأمـم
UNITED NATIONS	ET L'AGRICULTURE	Y LA ALIMENTACION	المتحـدة

Regional Office for Asia and the Pacific
Maliwan Mansion, 39 Phra Atit Rd
Bangkok 10200, Thailand

Tel. (662) 697-4000
Facsimile (662) 697-4445
E-Mail Address: FAO-RAP@FAO.ORG

Our Ref: IL 38/104

Your Ref:

0 3 OCT 2005

Dear Dr. Chaudhary,

High Level Policy Dialogue on Biotechnology for Food Security and Poverty Alleviation: Opportunities and Challenges

Organized by APAARI, FAO and GFAR

Bangkok, Thailand, 7-9 November 2005

Further to the initial information on the above mentioned policy dialogue which you may have received during the first two weeks of August 2005 from the Executive Secretary of APAARI, we are pleased to cordially invite you to participate in this meeting. This policy dialogue is being jointly organized by the Asia-Pacific Association of Agricultural Research Institutions (APAARI), the Food and Agriculture Organization of the United Nations (FAO), and the Global Forum on Agricultural Research (GFAR).

The main objective of this policy dialogue is to facilitate appropriate policy decisions by developing countries of Asia and the Pacific region with respect to application of biotechnology in their food and agriculture sector in addressing poverty and hunger, in accord with the World Food Summit and Millennium Development Goals. The policy dialogue is expected to promote greater understanding of the issues, sharing of knowledge on new developments and findings, and raise awareness of their potential benefits and risks and the implications in terms of needed regulatory framework, institutional capacity building and human resources development.

In the above context, your participation will be very important to attain the objective of the policy dialogue. The Concept Note and Provisional Agenda are attached herewith for your kind information. Other background documents, detailed information note, and final agenda of the policy dialogue will be sent to you in due course.

Dr. Krishan Bir Chaudhary
A-1, Nizamuddin West, New Delhi 110013
India

Kindly confirm your participation as early as possible so that we can reserve your accommodation in the hotel where the meeting will be held. Expenses for your travel and per diem will be met by APAARI/FAO/GFAR. In this regard, you may wish to purchase the air ticket (least cost economy class travel by the most direct route) and have the cost reimbursed at the venue of the policy dialogue, in US dollar equivalent based on original receipt for the purchase of the air ticket and the copy of the air ticket.

Kindly send the confirmation of your participation, travel schedule and personal details in the enclosed form to Mr. P.K. Saha, Liaison Officer, APAARI, c/o FAO Regional Office for Asia and the Pacific (FAO-RAP), 39 Phra Atit Road, Bangkok 10200, Thailand, Tel: 66-2-6974373, Mobile: 01-9188189; fax: 66-2-6974408; E-mail: pksaha@apaari.org and apaari@apaari.org, with a copy to Dr. Purushottam K. Mudbhary, Senior Policy Officer, FAO-RAP, at his e-mail: Purushottam.Mudbhary@fao.org.

We look forward to receiving your confirmation soon and to your active participation in the policy dialogue.

With best regards,

Yours sincerely,

He Changchui
Assistant Director-General and
Regional Representative
FAO Regional Office for Asia
and the Pacific (FAO-RAP)

Raj Paroda
Executive Secretary
Asia-Pacific Association
of Agricultural Research
Institutions (APAARI)

United Nations Conference on
Trade And Development

Government of India
Ministry of Commerce And Industry

August 16, 2005

Dear Dr. Krishan Bir Chaudhary,

Subject: National Seminar on TRIPS- CBD and Subsidy Issues at the WTO - 25 August 2005, New Delhi.

We are pleased to inform you that the Ministry of Commerce and Industry, Department of Commerce, Government of India, and UNCTAD are jointly organizing a National Seminar on TRIPS- Convention on Biological Diversity (CBD) and Subsidy Issues at the WTO to be held in New Delhi on 25 August 2005. The Seminar is being organized in the background of the forthcoming Hong Kong Ministerial Meeting of the WTO scheduled to be held in December 2005. **The venue for the meeting is Taj Mahal Hotel (Longchamp), Man Singh Road, New Delhi. Mr. Kamal Nath, Hon'ble Minister for Commerce and Industry is expected to deliver the keynote address during the National Seminar. The draft agenda for the meeting and the issue paper are attached.**

The National Seminar will discuss issues concerning relationship between the TRIPS Agreement and the Convention on Biological Diversity, the protection of traditional knowledge and folklore. Finding a solution for implementing the TRIPS Agreement and the CBD in a mutually consistent manner has been a matter of particular interest to India.

Discussions will also be held on certain aspects of the Agreement on Subsidies and Countervailing Measures (inclusion of capital goods in the definition of "inputs consumed" and issues relating to export competitiveness). We intend to bring together representatives of trade and industry, civil society, consumer organizations, legal experts and policy makers involved with this field.

We would value your participation at the meeting and would be pleased if you could contribute to enriching its deliberations. The final agenda of the meeting and related documents would be sent to you on receipt of your confirmation to participate in the meeting. We would also be thankful if you could circulate this invitation to other stakeholders who may have an interest in participating in the workshop or could provide inputs for the meeting. **Please confirm your participation in the workshop (giving all particulars such as Name, Designation, Organisation, Email Address, Postal Address, Fax Number, Phone Number and brief Profile/Bio-Data) to Ms. Celine Fernandes, UNCTAD, Email: celinefernandes@unctadindia.org , Ph: +91-11-24635036/54/55 (Extn.11) , Fax: +91-11-24635000.**

With kind regards,

Mr. G.K. Pillai
Additional Secretary
Department of Commerce

Dr. Veena Jha
Coordinator
UNCTAD India Programme

No 600/CP/NAC/05

SONIA GANDHI
CHAIRPERSON
NATIONAL ADVISORY COUNCIL

2, MOTI LAL NEHRU PLACE
NEW DELHI - 110 011
PHONES : 011-2301 8669
011-2301 8654
FAX : 011-2301 8646

April 1, 2005

Dear Shri Chaudhary,

I have received your letter dated 21st March, 2005 alongwith enclosures regarding the National Seed Bill-2004.

My office is requesting the concerned Ministry to look into the issues raised by you.

With good wishes,

Yours sincerely,

Dr. Krishan Bir Chaudhary,
A-1, Nizamuddin West,
New Delhi – 110 013.

Amit Agrawal
Deputy Secretary
Tel. 23012613

प्रधान मंत्री कार्यालय
नई दिल्ली - 110 011
PRIME MINISTER'S OFFICE
New Delhi - 110 011

D.O. No.300/31/c/2/2004 ES I   November 1, 2004

Dear Shri Chaudhary jee,

I am directed to acknowledge the receipt of your letter dated 18.10.2004 addressed to the Prime Minister regarding the revocation of Illegal Patent of Indian Wheat Strain by European Patent Office, Munich, (Germany).

With regards,

Yours sincerely,

(Amit Agrawal)

Dr. Krishan Bir Chaudhary
A-1, Nizamuddin West,
New Delhi 110 013.

कृषि और सहकारिता विभाग
कृषि मंत्रालय
कृषि भवन, नई दिल्ली-110001
. Radha Singh

Department of Agriculture & Cooperation
Ministry of Agriculture
Krishi Bhavan, New Delhi-110001

सचिव, भारत सरकार
SECRETARY
Government of India

<u>D.O.No.20/Secy(A&C)/2004</u>   June 24, 2004

Dear ~~Shri~~ Choudhary ,

    Hon'ble Prime Minister has desired to meet with the representatives of the farming community to understand their problems first-hand.

    For this purpose, you are cordially invited to attend the meeting on Monday, the 28th June, 2004 at 12 Noon at 'Panchvati', 7, Race Course Road, New Delhi. Your travel costs will be reimbursed by the Ministry on submission of claim.

    For security reasons, we would appreciate your reaching 7, Race Course Road at 11:00 AM. Arrangements for entry passes and other formalities would be made by us.

With regards,

Yours sincerely,

(Radha Singh)

Shri Krishan Bir Choudhary,
A-1, Nizamuddin West,
New Delhi

88th SESSION OF INDIAN SCIENCE CONGRESS
"Food, Nutrition and Environmental Security"
January 3 - 7, 2001

Dr. R. S. Paroda
General President

NAAS/ISC/3/2000/ 770
December 4, 2000

Dear Shri Chaudhary,

It gives me great pleasure to inform you that the 88th Session of the Indian Science Congress will be held at the Indian Agricultural Research Institute (Pusa), New Delhi, from 3 to 7 January 2001. The theme of the Congress is "**Food, Nutrition and Environmental Security**". This is an event of great national significance and is always inaugurated by the Prime Minister. More than 5000 participants from India and abroad are expected to take part in this Congress.

A very elaborate scientific programme covering various facets of science has, been drawn on this occasion. Also an important feature of this congress pertains to organization of a Public Forum for the *Scientists, Farmers and Students* where all panelists will speak on the Theme of the Congress.

In view of your significant contributions in the field of innovative farming and experience of Indian agriculture, I have great pleasure in inviting you to participate in the Public Forum on the Theme of the Congress, to be held on *04 January 2001 at 16.00 – 18.00 hrs.* and share your views briefly in 5-7 minutes on the subject of the Congress and also your own perception as to how Indian science could help the farmers in future.

I keenly look forward to your active participation and valuable inputs in the deliberations. Kindly do confirm by the return of post as time left to finalize the programme is very short. A copy of the programme is also enclosed herewith.

With my best regards,

Yours sincerely,

(R.S. Paroda)

Shri Krishan Bir Chaudhary
A-1, Nizamuddin (W)
New Delhi

Lawickse Allee 11
Wageningen

Postbus 88 > P.O. Box 88
6700 AB Wageningen
Nederland > The Netherlands

Tel +31 317 49 01 11
Fax +31 317 41 85 52

Telex 45888 intas nl
E-Mail IAC @ IAC.AGRO.NL

IAC ~ Internationaal Agrarisch Centrum
International Agricultural Centre

Dr Krishan Bir Chaudhary
Chairman of the State Farms Corporation of India
"Farm Bhavan"
14-15 Nehru Place
New Delhi 110019

Our ref.: hjm/148   Your ref.:   Wageningen, January 4, 1996
Tel: +31 317 490295

Dear Dr Chaudhary,

Thank you for your letters of November and December 1995 with regard to your company's visit to
the Netherlands in October.

I have rang with Incotec and send them a copy of your letter, which they would follow up. I gather that you
will receive the required information from them in the course of this month.

If you meet any of the other delegation members please give them my regards. It was nice travelling for me
in their company.

Kind regards to you and your family and best wishes for 1996, also on behalf of our director Drs Van de
Weg,

Yours sincerely,

INTERNATIONAL TRAINING

H.J. Manting

Centre Agricole International   Centro Agricola Internacional

2352
डी. ओ. नं. ....ट्टना सं./प९१०९ नं.आई.पी.(बाबी)
D. O. No ....MOS/(A&C)/95-VIP(I)

ARVIND NETAM

कृषि और सहकारिता राज्य मंत्री
भारत
नई दिल्ली-110 001
MINISTER OF STATE
FOR AGRICULTURE & COOPERATION
INDIA
NEW DELHI - 110 001

19th Dec.,1995

Dear Chandhary.

    I acknowledge with thanks, your letter dated Dec
01, 1995 alongwith the Report of your team's visit to
Switzerland and Netherlands under FAO's sponsored study
tour programme under NSP-III. I am sure, the knowledge
gained by such visits would enrich our current practice
and pattern concerning the seeds.

    Kindly let me know the Team's recommendations and
areas of interest which, if adopted, could contribute
further to the development of seeds in our country.

With regards

Yours sincerely,

(Arvind Netam)

Dr Krishan Bir Chaudhary,
Chairman,
State Farms Corporation of India Ltd,
Farm Bhavan,
14-15, Nehru Place,
New Delhi 110 019.

Comments

It is very refreshing and invigorating to learn from Dr Krishan Bir Chaudhary's book of collected essays that all is not lost in the agriculture sector.

The fresh and grassroots based insights particularly on the most burning issues, genetic modification as the only agricultural technology for ending the Indian farm crisis, has been rightfully questioned and rejected as it sounds the death knell of the marginal and small farmers in the country. Similar is the case when Dr Chaudhary identifies major pitfalls in WTO or FTA compatible trade agreements.

There is indeed a glimmer of hope in the saner advice carried between the covers. The short and crisp narratives are certainly most apt for those feigning 'know-how' and 'busybodies'.

It is a well known fact that the Indian policy makers at the top have been missing the wood for the tree. The book in every page reminds our Prime Minister Dr Manmohan Singh that he has been deviating far from his first principle of the rainbow principles that he gave from the ramparts of the Red Fort way back in 2004 on the Independence Day.

The alternatives suggested are easy to implement but requires sensitivities and resilience of the Indian farmers be it GM, climate change or failure of oligarchic market formulations.

This book must be made a compulsory reading at the top in the Krishi Bhavan, Agricultural Universities as well as the Planning Commission. I will recommend a special tutorial based on the book for the agricultural price commission officials as well as those charged with policy formulation.'

Prof. J. George [Ph d. Economics, Delhi School of Economics]
Chief Promoter-Strategic Economic Management Initiative in
Governance (SEMIG) Borgo Santa Croce 4, Firenze-50122 (Italy)

Dr. Krishan Bir Chaudhary has over the years developed with his articles into India's foremost agriculture analyst. The articles in the book are an excellent critical coverage of government policies and issues concerning Indian farmers at the domestic and International level.

The book "Development Misplaced", which is a collection of author's works, adequately reflects the current state of agrarian distress and situation leading to farmers' suicides.

I sincerely hope the policymakers will take lessons from suggestions incorporated in this book – Development Misplaced – and formulate an alternative development strategy

S.P.Ghulati,
Former Secretary,
Government of India

The book on Major Challenges in Indian Agriculture is a collection of papers by Shri Krishan Bir Chaudhary over the past few years. The book deals with almost all important issues, such as (i) farmers' rights on seeds, (ii) dangers of genetically modified crops and MNCs, (iii) foreign direct investment in retails, (iv) impact of WTO, (v) Privatisation of markets, and (vii) questions relating to agricultural productivity improvement, food security, farm distress, need for farmers' unity to change policies etc. I am sure that the book will make an interesting and useful reading by all concerned, including India's Planners and Policy makers and researchers.

Dr.T. Haque
Former Chairman,
Commission for Agriculture Costs & Prices,
Govt. of India